The Laws of Eternity

# The Laws of Eternity

*Unfolding the Secrets
of the Multi-Dimensional Universe*

---

## Ryuho Okawa

Lantern Books • New York
A Division of Booklight Inc.

Lantern Books
One Union Square West, Suite 2001
New York, NY 10003

© Ryuho Okawa 1997

English Translation © Ryuho Okawa 2001
Translated by The Institute for Research in Human Happiness Ltd.
Original title: *Eien-no-Ho*.

Library of Congress Cataloging-in-Publication Data

Okawa, Ryuho, 1956–
 [Eien-no-Ho. English]
 The laws of eternity : unfolding the secrets of the
multi-dimensional universe / Ryuho Okawa.
   p. cm.
 ISBN 1-930051-63-8 (alk. paper)
 1. Kofuku no Kagaku (Organization)—Doctrines. I. Title.
 BP605.K55 O3213 2001
 291.4—dc21

2001050190

# Table of Contents

## Chapter Three: The World of the Sixth Dimension

## Chapter Four: The World of the Seventh Dimension

# Preface

As the name claims, the contents of this book really are the "The Laws of Eternity." The Eternal Truth that has never been preached before, and will never be taught again, is explained logically and condensed into this one book.

The three unique characteristics of the Law I teach are: a vast structure of the Law that encompasses all aspects of the Truth of life, a theory of time, including a description of the historical roles of the high spirits—the Nyorai and the Bosatsu—and a theory of space, which presents a concise explanation of the multi-dimensional Real World, the world we go to after death.

These three characteristics are represented in the trilogy of books: *The Laws of the Sun* (the structure of the Law) and *The Golden Laws* (theory of time) have already been published, and this book, *The Laws of Eternity* (theory of space). Taken together, these three works are an outline of the Law of El Cantare.

This book throws light on the eternal mysteries, and it reveals the ultimate secrets of the terrestrial spirit group, hitherto hidden behind a veil of myth.

I send this book out to you with the earnest hope that we can surmount all difficulties and work toward the unification of all the religions of the world.

*Ryuho Okawa*
*July 1997, for Japanese edition*

# One: The World of the Fourth Dimension

## 1. This World and the Other World

"Where did I come from and where will I go when I die?" This is a vital question that lingers somewhere in our minds, but it is one to which very few people have been able to offer a clear answer. The reason for this is that in order to provide an answer it is necessary to elucidate upon the relationship between this world and the other world. Unfortunately, as things stand at present, there is neither enough accumulated knowledge on the subject nor an established methodology capable of producing an explanation.

The only clues we can find, faint as they may be, lie in the activities of the psychics who have appeared on Earth through the ages. However, there are many types of psychics, and although some of them can be trusted, the vast majority have slightly unbalanced personalities, which is why the people of this world are unable to believe their word. For instance, a psychic may say that he has spoken with a particular spirit who told him that a certain event would happen in one year's time, but there is no way of proving what he says. This creates a feeling of uncertainty and makes it very difficult to trust the psychic's word. The same is true when it comes to an explanation of this world and the next; the root of this uncertainty lies in our inability to duplicate the experiences of the psychic.

If we could all share the same psychic experiences, nobody would doubt the existence of the other world. But, unfortunately, this ability is limited to only a few, special people. As a result, the majority of people do not know of the existence of the other world. Practical people do not want to accept the existence of the other world or the relationship between it and the world we currently live in.

People are forever pondering the meaning of life and the reason they exist. These are vital questions and are impossible to understand until we can grasp some idea of what kind of existence we represent in the vastness of the universe. If, as materialists believe, life begins all of a sudden inside the mother's womb, continues for sixty or seventy years before ending, and the body is disposed of at the crematorium, then we should live our lives accordingly. However, if, as men of religion would have it, there is another world from which souls come to be born into this world, where they live for several decades before graduating and returning to the other world where they will strive to improve further, then we require a different perspective on life.

If we compare life to education, materialists who claim that we only live once would be equivalent to those who say that education is over after the compulsory period of six years. They only see life in the limited perspective of elementary school education. On the other hand, those who believe that the spirit world exists and that human beings live eternally, going through the cycles of reincarnation, can look at life as a continuous course of education: once we have finished primary school, there is junior high, senior high, university, graduate school, and after that still a myriad of other things to learn.

If we look at these two instances, it becomes obvious that it is only from the perspective of eternal "education" that human beings can progress.

Those who believe that their life represents a single, brief sojourn in the world, flaring up like a match until it is consumed, are unlikely to discover the importance or the meaning of their time here. They will devote themselves to pleasure, to materialism and decadence, thinking only of themselves. It is only to be expected that if they believe they will only exist for a few decades, they should feel they would be losing out if they did not enjoy themselves to the maximum. On the other hand, however, people who believe in eternal life are able to know that services rendered to others will surely come back to them some day as nourishment for the soul.

In this way, it can be seen that having the perspective of this world and the other world is vital when considering our outlook on life, its purposes and mission. Without it, it is impossible to understand the true meaning of life and of human beings.

## 2. The World after Death

We refer to the place that people go after being separated from their physical bodies as the "other world," but what kind of a place is it? What kind of a world awaits us after death? Because they do not know what to expect, people become anxious and afraid, expressing their attachment to their earthly lives in the words, "I don't want to die." Ninety-nine people out of a hundred will say they don't want to die, but this is not simply because they find life in this world easy; it is the fear and anxiety about the next world that makes them feel anxious and afraid. There are some people, however, who find life in this world so unbearable that they overcome their fear and unease of the unknown, by choosing to take their own lives and set out prematurely on the journey to the next world.

Whichever way people may feel, I believe their fears are based on their ignorance of the other world, of the world after death. The fact that the study of the other world has never been made into a science has resulted in distress for vast numbers of people, and it is for this reason that I have come to act as a pilot to guide them across. Navigating without charts can be a harrowing experience, but with the proper maps a voyage can be undertaken with a minimum of unease. If we know where we have come from and where we are going, what continent we are aiming for, so to speak, if we can understand the charts, we can make a safe journey.

As I have already stated in many books, our life is not something that can be counted in decades; it is not merely the life of our body, but of something that continues through this world and the next. However, when faced with it, nobody wants to accept death. Patients in hospitals say they do not wish to die, and doctors do everything in their power to support the life of the patients. However, if we could look at it from the other world, we would see that as a person approaches death, their guardian and guiding spirits and the angels are standing close by. They have already begun the preparations to guide the person who is nearing death.

When the end finally comes, the spirit moves out of the body. However, at first, the person does not realize what is happening, and feels as if there are two of them—one lying in the bed and the other able to move freely. When the freed spirit tries to communicate with the surrounding people, it finds that it is totally ignored. At the same time, it discovers that it can pass through walls and other material objects, a fact that it initially finds quite shocking. It still believes that the body lying in the bed is its true self and continues to hover over it; you can imagine the shock it feels when it sees the body being taken to the crematorium to be burned. Not knowing what to do next, the spirit remains in the vicinity of the crematorium. It has not been told what sort of a life awaits it and understandably it becomes very upset.

It is at this point that its personal guardian spirit appears and begins to explain what awaits it. However, having lived in this world for many decades, the spirit finds the explanation difficult to understand and despite all the efforts of its guardian spirit, it is not easily convinced. For this reason, it often remains on Earth for several weeks, listening to the explanations of its guardian spirit. As the Buddhist services for the dead on the seventh and forty-ninth days after the day of death suggest, a spirit is generally allowed to remain on Earth for between twenty and thirty days after death. During this period, the spirit receives instruction from the guardian or guiding spirits until it is prepared to return to the heavenly realm.

However, there are those whose attachment to this world is too great, for instance those who remain devoted to their children, father, mother, wife, husband, or even to their land, house, wealth, or business. They become what are known as "Earth-bound spirits," doomed to wander about in this world. These are what people know of as ghosts and they are beings who remain unaware of their existence as spirits.

### 3. Memories of a Physical Body
Most people suffer from shock and dismay when they first find themselves in the other world, but as time passes they become used to life there. They gradually realize what it means to live without a body. It may take twenty or thirty days for them to discover they can go without having to eat or drink. Eventually, they realize the futility of trying to communicate with the people

who remain on Earth, as they cannot be heard, and give up their efforts in this respect.

As they cast off their old habits, they gradually acquire a new way of looking at things—what can be described as a spiritual sense. They discover they can float on air, pass through solid objects, and travel vast distances instantaneously. People who are still attached to their earthly lives and want to say goodbye to a relative or visit a friend for one last time can travel hundreds of kilometers to see them in the blink of an eye. This ability is very exciting in the beginning, but as time passes it loses its novelty. People settle down to consider their new outlook on the world and their place in it, in much the same way as a child does when it enters primary school, and finds itself confronted with a new world. Usually, memories of life on Earth gradually fade away, but for some spirits the memories grow ever stronger.

At some point, the spirits who have been wandering around the three-dimensional phenomenal world are eventually led away by friends or relatives who had died before them or by their guardian spirits to a "reception center" in the fourth dimension where they can reflect on their lives on Earth.

The basis of the spirits' introspection is the mistakes they made when their life is looked at from a spiritual viewpoint. In other words, the spirits reflect on the extent their way of life was centered on the physical rather than the spiritual. As a result of their reflection, those who focused on material things without awakening to their spiritual nature take it upon themselves to go to the area of the fourth dimension that is known as hell, where they are inflicted with severe ordeals. On the other hand, those who did not live a spiritual life but who confess the error of their ways frankly and are truly repentant of their wrongdoing are led to the Astral Realm where harmonized spirits live.

In this way, our future lives in the other world are affected by the memories of our life on Earth. There is no omnipotent being who sits in judgment over us; rather it is our own conscience, our true nature as children of Buddha, that judges us. If we feel we are still lacking in spiritual discipline, we may decide, after discussing the matter with our guardian spirit, that we should go to hell for further training. Although each soul makes its own choice, after a soul has been in hell for a long time it forgets that it made the initial decision to go there itself and gradually begins to feel discontent.

However, the really vicious souls do not go through this process of reconsideration and are cast straight to hell immediately after their death. These are the spirits who try to lead others astray. On Earth we are familiar with the characters of gangsters and other violent people, and if you think of these spirits as being of the same kind, you can understand what they are like.

## 4. The Nature of Angels

These days, most people seem to have difficulty believing in the existence of angels. Even devout Christians who believe in angels would find it difficult to accept their actual existence. Christianity is based on belief in "the Father, the Son, and the Holy Spirit," and, while Christians can understand the Father and even the Son, they have a lot of trouble when it comes to comprehending the Holy Spirit.

Most people seem to think that angels and devils belong to the world of fairy tales and do not believe that they actually exist in the present day. The mere thought of them would be enough to bring a cynical smile to the faces of about ninety percent of people. However, angels and devils are not nursery tales. The reason they can be found in stories throughout history, both in the East and West, in advanced and developing countries, is that they really do exist. "Angel" is the generic name given to high spirits, but they are not all the same. I will go into the subject at greater length later on, but suffice it to say that spirits exist at various stages of development, and the ones from the upper part of the Realm of Light of the sixth dimension and above are what are referred to as angels. They are spirits who have achieved the level of "special-mission high spirits" or the more advanced Bosatsu and Nyorai.

The angels at the initial level of their training are responsible for saving people who have only just left Earth. They do not preach the Law, instead they are engaged in practical tasks of saving souls. They act as guides to the people who have just left this world, looking after and educating them. In total they number in the hundreds of millions. These angels appear to people in a form that will be understood according to their ideology, beliefs, and religion. Thus, Christians are generally looked after by angels from the Christian spirit group, who teach them accordingly, while Buddhists are guided by Bosatsu from the Buddhist group. In other words, the souls of people newly departed from this world are instructed by beings they find it easy to believe in.

However, this is not to say that angels limit themselves to the other world. Many of them are reincarnated on Earth every few hundred to a thousand years. They come here to discipline their souls and purify the world, and at the same time, to remind themselves how it feels to live a human life. If they remain in the other world for too long, they find it increasingly difficult to understand how an earthbound human thinks or how their mind works, so in order to be able to fulfill their role as teachers, it is necessary for them to come down to Earth periodically and refresh themselves in the ways of Earth dwellers. An added advantage of their coming to Earth in this way is that, having regained an understanding of human sensibility, they are able to preach the Law more effectively and lead even more people to salvation.

In this way, the first thing that people experience after death is the work of angels. They are bright, luminous souls and when they come to a Christian they will appear with wings, to Buddhists they take the garb of a Buddhist priest, while to the believers of Shinto they appear in the robes of a Shinto priest. No matter whether the being is an angel or a high spirit, it will appear as a shining figure with a dazzling halo and, even if the beholder professes no religion, they will not hesitate to put his hands together in prayer. This is because human beings have an instinctive belief in high spirits and divinities.

### 5. A New Departure

People who have left Earth receive instruction from the angels, and gradually ready themselves to make a new departure. But what is meant by "a new departure"? It is a totally new experience. Of course, when we were born here on Earth we made a new beginning. No matter how great or how immature a spirit may be, once it is conceived in its mother's womb and born to this world, its memories are erased from its conscious mind; it experiences a new beginning.

In the same way, this new experience I talk of means that after spending several decades studying on Earth, the soul graduates from this life and moves on to a new school. In other words, the soul makes a new departure and encounters new teachers, new textbooks, and new lessons. First of all, these returnees to the fourth dimension are put through an intensive course on the meaning of spirituality.

In addition to receiving instruction from angels, new arrivals are instructed by old friends and teachers who have arrived in the other world ahead of them, who help them to accept and understand this new start. This provides them with important guidance for their future life in the other world. Although many of them will forget what they learned after they have settled down to their new lives, it helps them over the first stage of their transition.

Among the souls that arrive in the other world, there are those whose new departure leads them to hell. However, what I would like to make quite clear to all of you who are living on Earth is that hell is not a realm equal in size to heaven. The world in which we live is in the third dimension; while the other world stretches from the fourth to the ninth, tenth dimensions or beyond, hell, on the other hand, occupies only a tiny section of all this. It is just a lair of negative thought energy that lurks in a corner of the fourth dimension and in no way is it equal to the rest of the other world.

There are all kinds of people living in our world, but the sick are not a separate race so we have hospitals to look after them. In the same way, there are souls in the other world who are sick in spirit, and hell is where they go to discipline themselves and undergo rehabilitation. Although they may be sick in spirit and mind, they are struggling to achieve something in their own way. A healthy person can be taught to drive a car, ride a bicycle, to jump, or to run in long and short distance races, but a sick person may not be capable of any of these things. A sick person has to learn how to use crutches, or to walk with somebody supporting them by the arm.

When a soul arrives in the Astral Realm, which is also in the fourth dimension, it sees various kinds of beings it had never seen before, many of which have been mentioned in ancient tales on Earth. Dragons and water sprites do not live in this world on Earth, but they do live in the fourth dimension just as do small fairy-like beings that fly around the flower gardens. There is a vast variety of strange life forms in the next world, and in watching them newcomers gradually get adjusted to the new world.

### 6. The Essence of the Spirit

Now, I would like to talk about the true essence of the spirit. I have already explained that immediately after it has left the body, the spirit cannot accustom itself to its new spiritual existence. Shortly after leaving the Earth, it experiences a lot of difficulties in acquiring spiritual sensitivity. For instance, while the spirit was living on Earth, it had only to reach out and grasp something, but once it is in the spirit it is unable to hold anything, a fact that it finds it very difficult to come to terms with. Before long, however, a spirit begins to take its new sensitivity for granted, even though it may still not have realized that it is dead. Then it has to decide whether it should take the road that leads to heaven or that which leads to hell.

In making this decision, the important thing is to have a good understanding of what kind of person you have been, to realize what your essential nature is. This is what decides the kind of life you will lead in the next world. Even if you do not believe in the spiritual world, you should have learned about the other world from picture books, old stories or novels and such like. The problem is that we have no way of telling how true any of this information really is.

What kind of life will allow us entrance to heaven and what kind of life will lead us to hell? Today there are very few places one may turn to for an answer. Even if we do believe in the existence of the other world, there is nothing to tell us whether our deeds, when seen from a spiritual viewpoint, prepare us for heaven or for hell. The simplest way of assessing this is to refer to religious commandments to see if our actions are sinful or not, as we are told that "Sinful people go to hell whereas those whose sins are few may go to heaven." This way of thinking has guided humankind from time immemorial. East, west, north or south, this concept has been common to all cultures for thousands of years, the most famous religious commandments being the "Ten Commandments" of Moses and the "Code of Hammurabi" in Mesopotamia. The origins of these commandments are the teachings of the Guiding Spirits of Light. Their total message is too difficult to be understood by the average person and for this reason the message is presented in the form of commandments: "You may do this, but you must not do that." As a result, the easiest way for people, even those who are spiritually aware, to understand heaven and hell is to ask themselves whether or not they have lived their lives in accordance with the commandments.

The good thing about the commandments is that they provide a simple, readily understandable guide as to what is right and wrong. For instance, the most basic commandment is "Thou shalt not kill." This means that if you kill somebody you will go to hell, and if you do not, you might go to heaven. Another example is "Thou shalt not steal," which means that if you do steal you will go to hell, and if you do not, you might go to heaven. This dichotomy may seem a bit simplistic but nevertheless, the Light of Truth shines through it.

However, it is not the commandments that divide heaven and hell. The truth is that those who expressed their true nature as children of Buddha during their life on Earth will be able to ascend to heaven, and the more fully they manifest their true nature the higher they will rise. If, on the other hand, they do not awaken to their true nature as children of Buddha, and fail to encourage this true nature to manifest while on Earth, they will be inflicted with severe ordeals in hell.

## 7. The Unknown

I talk of hell, but when people are actually faced with it, nothing they may have read about it during their lives on Earth can prepare them for the terrible shock that its reality will present them with. They may have read stories about there being several types of hell, some of them inhabited by demons and devils. However, when they are actually faced by these creatures, their horror will know no bounds. Some of them appear to be three to four meters (ten to thirteen feet) tall, while others seem to be attacking with knives in their hands.

In the Hell of Lust people writhe in a sea of blood, while in the Hell of Hungry Spirits people of skin and bone, like the victims of a famine, stagger around calling out for something to eat. In the Hell of Beasts people lose their human form. The Japanese writer, Ryunosuke Akutagawa (1892-1927), wrote about people who fell to the Hell of Beasts, describing the way in which some would have the body of a horse, ox or pig but with the human face; this depiction was quite accurate. There are even those who take on the body of a snake and are forced to slither along the floor of hell.

None of them know why they are in the state they are, but it is because they are unaware of their spiritual nature. In the spirit world, you become what you really are. They did not know that their secret thoughts were already materialized in the world within. They believed that just because their thoughts could not be seen from the outside, they could think whatever they liked. If their thoughts had been visible to those around them while they were living in the flesh, they would have been ashamed to appear in public. Therefore it is easy to imagine the shock they feel when they return to the other world to find that their thoughts are transparent to others, and that they change into a shape that conforms with what they think.

If people in this world on Earth who are filled with envy and spite turned into snakes, they would quickly realize that they had mistaken thoughts. But since this does not happen under the laws of this physical world, people remain unaware of their mistakes. However, in the other world thoughts are instantly actualized. For instance, a person whose thoughts are only filled with desire for the opposite sex will enter the Hell of Lust where they spend their time in search of sexual satisfaction. Sly people who spend their time trying to cheat others will find themselves in the shape of a fox in the other world. People who are eaten up with resentment and hate will awaken to find their bodies have turned into snakes. These are only a few examples of the numerous animal forms that can result from wrong thoughts.

These animal-like human spirits will seek temporary respite from the agonies of hell by possessing people still living on Earth. However, they cannot possess just anybody they like, it has to be a person who has already created a hell within. Living people create various worlds of thought within their minds, and those who have chosen to create a hell are vulnerable to possession by the spirits of hell. People who have created a hell of lust within their minds will be visited by the denizens of the Hell of Lust. People who have created a hell of beasts in their minds can expect to be possessed by animal-like spirits. People who have embraced the abysmal hell within or who are so diseased philosophically and religiously that they try to lead others astray will be invaded by the fallen men of religion or thinkers who now inhabit the Abysmal Hell. It is because living people create these hells within them that the spirits of evil are able to possess them.

## 8. Eternal Life

All the spirits suffering in hell repeat the same refrain, which is, "I would rather die than continue my life like this." They abuse

Buddha and God in every way possible and say, "It would be better if you had killed me straight out instead of making me live like a snake," or "Why don't you kill me? Anything would be better than trying to stay afloat in this sea of blood." The inhabitants of the Abysmal Hell, trapped in darkness, or in the desert, or sealed into caves, are constantly saying, "If this is all the life that remains to me, why didn't you put an end to it?" Through the use of my second sight, I have seen men who once led vast religions and were praised by one and all suffering alone in the infinite darkness of the deep swamps of the Abysmal Hell. Some of them are founders of religions which are now in their second or third generations, and the question they all ask is, "Why am I in a place like this when once I led tens of thousands, even millions of followers? I would rather my life were ended than to spend the rest of it in a place like this." This is because they do not know their future; trapped in the pitch blackness they have no idea how much longer they must suffer.

Souls are for ever. Souls have eternal life, and for people who have lived in harmony, whose hearts are pure, and who have returned to heaven after death, eternal life is the ultimate blessing, for they can continue their life in that wonderful place. However, for people who go to hell, eternal life is itself a punishment and castigation. If they could only die, they would no longer have to suffer the torments of hell, but life never ends and this very fact serves to afflict them.

If people could understand the nature of the Real World while they were still in this life they would realize that it did not pay to think or do evil. However, they do not believe in eternal life and feel that as they are only here for this life they can do as they please. They do not care how many people they may hurt through their actions as long as they can become rich and powerful. If only they knew that this would win them eternal agony instead of the enjoyment of eternal life, they would realize just how shortsighted they had been.

If only people could know that a life of goodness, no matter how modest, will lead to unparalleled bliss, they would ask themselves why they did not do more good while they were on Earth. One act of kindness in this world is repaid tenfold in the next. Life in this world is very difficult, but if we manage to live a heavenly life while groping our way forward in the darkness, it is worth five or ten times of the spiritual discipline in the other world.

If we think that a small misdeed will not make any real difference and repeatedly do things we know to be wrong, we will find that this too will be repaid five or tenfold. This is the underlying severity of the world. Therefore, you should not do good merely because people will praise you for it, nor should you refrain from doing evil just because it would be morally wrong. If you really want to look after yourself, you will realize you have no choice, you cannot afford to do evil and must devote yourself to good. Once you acquire knowledge of the true perspective of life, the true perception of the world, you cannot do otherwise. It does not pay.

The people who end up in hell really hate to put themselves at a disadvantage, more than anybody else, but they do not realize that their actions do just that. For this reason it is important that we awaken them to the Truth as soon as possible.

## 9. The Memories of Past Lives

I have told you various facts about heaven and hell, but the most surprising thing about the fourth dimension is that when we are there we recall the memories of our previous lives. This is the most interesting and exciting experience that awaits us. In this world we progress from infancy, through the educational system where we learn what we will need for adulthood, and eventually to old age, taking pride in the experience we have gained through the course of our long lives. However, when we return to the next world, we realize that our true lifetime experience is much, much longer. It is not to be measured in thousands or tens of thousands of years, but rather in millions. All of us contain in our souls the memories of tens of millions, even hundreds of millions of years, and when we return to the Real World these memories come flooding back. We realize that we have been living as a human for all that time.

However, the people living in hell find it difficult to recall their past lives. Life in hell is a painful, grueling experience. When one is in great agony it is difficult to reflect on the past. For instance, if a person who is suffering from a severe toothache is told to recall the past and consider their lives, they will be too distracted to do so. The same is true of the spirits that are writhing in the agonies of hell. Although they are theoretically capable of recalling all their previous lives, they are in no practical condition to do so.

On the other hand, those who have returned to heaven can readily recall their previous lives, although the extent to which they can achieve this varies according to individual ability. Those who had lived on Earth as ordinary people will be able to remember one or two previous lives, but it will not be a clear memory. They will simply remember having done something a long, long time ago. However, as they develop into higher spirits they will find that they can remember their past ever more vividly.

A Bosatsu, or high spirit of the seventh dimension, can remember its lives back for several tens of thousands of years while a Nyorai of the eighth dimension can remember even further into the past, and, if they put their mind to it, they can recall things that happened hundreds of thousands, even millions of years ago. When a spirit reaches the level of the Grand Nyorai of the ninth dimension, they can recall even the details of the creation. They know how they came into existence hundreds of millions of years ago, how the Earth was formed and how humankind has evolved since then. All the details remain clear in their minds. In this way, it can be seen that the ability to remember the past depends on the rank of the spirit. It is like climbing an observation tower.

If you climb a high tower, it is possible to see vast distances, but if the observation deck is low you can only see the nearby scenery, and if you descend into a basement it is impossible to see anything at all. If you are in the basement, that is to say, hell, you will see nothing, but the higher you climb the further the vista expands. In the same way, the higher your rank among the spirits the further back your memory can reach. So it can be seen that although you will remember your previous lives, the extent to which this can be achieved depends very much on the individual. Some people can only remember one previous life, others several, still others may recall several dozen, while some may be able to cast their minds back over hundreds of previous existences. It is a very mysterious process. But if you cultivate your spiritual awareness you will find that you are able to see the past, the present, and the future.

## 10. Evolution

So far I have explained various aspects of the fourth dimension, that is to say, what will happen to you after your death, but I am sure that many of you will be wondering why it should be that way. Why is there a heaven and hell? Why do we not know about them while we are still alive? Why do we have a body and a spirit? Why do we not live in spirit form in this world? These are just some of the questions that people may ask.

The transition from the body to the spirit can be likened to the metamorphosis of the cicada from the larval stage to the adult insect. After many years living under the ground, the cicada nymph climbs a tree, then, clinging to the trunk, its skin splits and out climbs the adult. It stretches its wings and then is free to fly wherever it pleases. Similarly, the ugly green caterpillar crawling across a leaf will one day hide itself in a chrysalis and later emerge as a beautiful butterfly. The life cycle of the butterfly was created by Buddha with the express desire to demonstrate to us the cycle of reincarnation that we ourselves experience, and to demonstrate that, in the same way that the insect changes its form, we also change as we evolve.

It may seem strange that a caterpillar should turn into a butterfly, but nobody can deny that Buddha created it in that way. It is hard to believe, watching the grotesque creature with its dozens of stubby legs gorging itself on leaves, that it will one day grow wings and fly freely through the air in a blaze of color, but it provides us with an intimation of the spiritual evolution of humankind.

Why did Buddha make these insects in this way? It is an example of His compassion. He could just as easily have made butterflies free to fly from birth, but by showing us the limitations of a ground-bound life, He enabled us to appreciate the freedom of flight. I am sure that there are very few human beings who wish they could have been born as butterflies, but there is a splendor in their flight that we can never understand. Butterflies probably know a happiness that we cannot share and this too shows us Buddha's compassion.

In the same way, people are forced to live limited lives while contained within physical bodies, but eventually they are able to cast the body aside and return to their true form as spirits. When this happens they are able to appreciate the true splendor of their being so much more. While living on Earth, we are unable to achieve everything we want, which leads to impatience, exhaustion, and a feeling of helplessness, but when we arrive in the Real World, our every thought will manifest instantly. When people realize the marvel that this represents, they feel that life in the Real World is immeasurably better than what they had experienced on Earth.

Spiritual evolution is part of the very structure of the world and was prepared for us by Buddha. We need to experience this metamorphosis, this sublimation to the next stage, to understand the true meaning of happiness. Being spiritual means to share in the essence of Buddha, and this is an experience that every human being is able to undergo. This is truly wonderful, and we live in such a marvelous world.

Even if we are obliged to experience life in hell for a hundred, even two hundred years, in the long term we will realize that this is not necessarily a bad thing and can be considered a whetstone with which we polish ourselves; in the severe ordeals of hell we have no choice but to face the greatest faults in ourselves and work to correct them. It is possible to regard the experience of hell as a stage in evolution.

This is not to say, however, that spirits should be left in hell unaided. While they are there, they do experience great pain, and in order to deliver them from this suffering they must be shown the error of their ways and led in the correct direction. Only in this way can Buddha's will be truly done. People struggle to correct their faults through the help of others. But hell exists to provide them with an opportunity to realize where they have gone wrong even without outside aid. It may seem like a backward step, but if you look at it with a long-term perspective, you will realize that it is all part of evolution.

# Two: The World of the Fifth Dimension

## 1. The World of Goodness

In Chapter One I explained the mysteries of the world we first enter after casting off our human bodies and returning to our spiritual form, and in this chapter I would like to discuss the next dimension. Some of us may accept that our world is part of a multi-layered structure consisting of the third, fourth, fifth, sixth, seventh, eighth, and ninth dimensions. The third dimension is enveloped by the fourth, the fourth by the fifth, and so on—each progressive dimension encompassing those below it to create a vast, onion-like structure. This means that the people of the fourth dimension do not live in a totally separate world. The fourth dimension co-exists with the third and exerts various influences over it while the fifth dimension exists over the fourth. One interesting thing about this structure is that the inhabitants of the higher dimensions may influence the people on a lower plane, but not the other way round. The inhabitants of the fifth dimension are free to travel to the fourth dimension where they offer guidance to the inhabitants but, with very few exceptions, the inhabitants of the fourth dimension are unable to visit the fifth dimension.

This may be rather difficult to accept, but when you return to the other world you will find it is true. That there is a spiritual hierarchy in the other world is a fact accepted by Buddhism, mysticism, and theosophy, and is confirmed in numerous ancient manuscripts. These manuscripts say that the spiritual universe is not merely divided into two halves, this world and the other, but that the other world is itself divided into numerous different realms both horizontally and vertically. An eighteenth-century European psychic, Emanuel Swedenborg, made numerous visits to the spirit world and wrote down what he saw there. Among his records, he notes that, when you look up, there seems to be an invisible screen covering the sky and beyond that is what looks like another world. In actual fact there are no such visible layers, but the different worlds do exist, one above the other.

All this being said, what is the difference between the inhabitants of the fourth and fifth dimensions? The fourth dimension represents the first step in the spiritual world and its inhabitants may be compared to first-graders in elementary school. They are still not fully aware of the relationship between the body and the spirit, or the soul and matter. Their lives still remain a mixture of Earthly and spiritual customs. After living in the fourth dimension for a while, these "first grader" spirits begin to evolve—quick people requiring only a matter of days, slow people taking anything up to hundreds of years. When they are ready, angels or their guardian and guiding spirits come to lead them up to the fifth dimension.

You may ask what kind of world the fifth dimension is. In a word, it can be described as being the world of "goodness." The fifth dimension is the place where those who have chosen good and abandoned evil gather. The souls that inhabit the fifth dimension share a natural tendency toward good, and when living in this Earthly world had made efforts to choose good, turning their backs on evil. They have realized that goodness is the characteristic that the Divine looks for in humans.

## 2. A Spiritual Awakening

When I speak of the "goodness" of the fifth dimension, I do not merely mean goodness as contrary to evil, but as realizing one's nature as a child of Buddha. To put it another way, it refers to a person's spiritual awakening. In our lives on Earth, material and spiritual things co-exist side by side. We live in the midst of material things and spend our time worrying about making a living, earning a salary, and buying, using, and discarding material objects. If, however, a person can find the time in the evenings or at weekends to attain spiritual joy despite living in this materially centered world, they must be considered blessed indeed. Of course, there are some people who never feel any spiritual pleasure, devoting themselves to gambling and other pleasures of the flesh, but most of us do not find this sufficient to satisfy the soul, and feel a kind of nostalgia toward spiritual pursuits such as reading, music, or art. This is in fact a nostalgia for the world we originally came from.

The fifth dimension is often referred to as "the Realm of the Good" or "the Spiritual Realm," and this is because people who have awakened to spirituality gather there. The inhabitants of this realm are aware of themselves as spiritual existences to a remarkably high degree.

On the other hand, the people who live in the fourth dimension, which is known as the "Posthumous Realm," are all at different levels of personal enlightenment. Some of them still do not recognize that they are in fact spiritual beings while others may have a fairly good grasp of their spiritual state but still have not succeeded in attaining a complete understanding of spirituality. They still have not reached the stage of searching for good within the essence of the spirit, based on the knowledge that they are themselves spirits.

Once people reach the fifth dimension, however, they are quite aware that the essence of human beings is spirit, they are searching enthusiastically for good. They believe in religion, though it may only be a vague stirring of faith, but it is something they all share. They may believe in God or Buddha, and their personal philosophies will reflect their religious background, but they all share a basic belief in some form of deity. They feel that the Divine is close at hand, and live their lives for its sake.

Many people living in the fifth dimension continue to involve themselves in the same work they had undertaken on Earth. There are carpenters and merchants who supply the needs of the people and numerous other professions that belong to what we on Earth refer to as the service industries. In this way, many of them continue in their professions, although in the fifth dimension they do not do it for money; rather they feel that by so doing they please the Divine and this in itself affords them pleasure.

### 3. The Joy of the Soul

The inhabitants of the fourth dimension cannot fully appreciate the joy of the soul. Instead they feel what is known as *surprise* of the soul. In other words, they are still experiencing the surprise and novelty of finding that they are, in fact, souls. They are amazed by the numerous strange things they can experience in their new state as souls. But the people of the Realm of the Good of the fifth dimension can feel the true joy of the soul. The reason for this is that the people of the Posthumous Realm have yet to differentiate fully between life in the fourth dimension and life on Earth, still clothing themselves in a corporal garb known as the "astral body." However, when a person enters the fifth dimension, this is cast away and that person's soul becomes more refined. In the fifth dimension, the soul comes to the fore. By a soul, I refer to a spirit that has the memory of life as a human being and it is this soul that begins to feel joy. But what kind of joy does it feel? When does a soul feel joys? Basically there are two ways.

First, the soul feels joy when it feels its own self-improvement. You may ask when this happens, and the answer is when it is able to identify itself with goodness. The soul realizes this when it feels that it is helping other people and it fills it with great happiness. The same is true for us on Earth when somebody says to us, "I am so glad you are here," or "It is thanks to you that things went so well." It makes us very happy because when we know that we are able to help people in this way we are filled with the joy of self-expansion, the joy of self-improvement. We realize that we are not only living our lives for our own sake but that other people find happiness as a result of our existence. In other words, it is when our existence is worth more than a personal life that we have succeeded in self-expansion and self-improvement. Therefore, we can say that the first way in which we can feel the joy of the soul is when we are able to bring joy to others and realize our own goodness.

The second way in which we can experience the joy of the soul is when we acquire new knowledge. The knowledge I am talking about is not the kind that can be gained through the study of worldly subjects, but the discovery of something concerning the world that Buddha created. Every time we make such a discovery our soul is filled with joy.

The soul embodies numerous characteristics, talents, and powers, but the inhabitants of the fifth dimension are not aware of them all. There are some among them who still like to eat. Of course, they are quite aware that they do not have to eat to sustain their lives, but they feel that eating is one of the great pleasures of life, whereas there are others who find delight in the preparation of food. People who were farmers on Earth enjoy growing crops and spend their time in the Real World working in the fields. In this way, we find people in the fifth dimension working in all kinds of professions. But gradually they come to realize that this is not the true way in which they should live, that spiritual satisfaction can be found without having to work in this way.

Even if they think they are growing potatoes, these are not really potatoes in the three-dimensional meaning of the word, they are "spiritual" potatoes that have been created within people's minds. They gradually come to realize that the richer and fuller they become, the more beautiful are the potatoes they produce. On Earth there are all kinds of scientific methods of watering and fertilizing the crops, but this is not necessarily the same in the Real World. The more care, the more love that is applied to growing things, the more beautiful the result.

The inhabitants of the fifth dimension gradually become aware of the structure of the spiritual world where our very thoughts become facts. Thus, the acquisition of knowledge is translated into the second form of joy for them.

### 4. The Flow of Light

In the last section I discussed the two ways in which the soul feels joy. The first is when it is able to offer other people assistance and the second is when it manages to acquire new spiritual knowledge. In this section, I would like to talk about spiritual knowledge in greater depth. In the Posthumous Realm of the fourth dimension, inhabitants do not really understand that the true essence of a human being or of the soul is Light that originates in Buddha. Once a soul reaches the fifth dimension, however, it gradually becomes aware of the true nature of Light. It awakens to the intensity of Light and realizes that unlike an electric light or candle, the

Light of Buddha is a flow of real energy. It is a strange sensation. The inhabitants of the Realm of the Good in the fifth dimension become aware of the source of this energy. There is a giant sun in the sky, but this is different from the one we are used to seeing in the third dimension; it is called the Spiritual Sun.

Let us look now at the true nature of the Spiritual Sun. The sun that shines down on the third dimension is familiar to us all, and the Spiritual Sun is the spiritual form of this. Just as a soul resides inside a human body, a great soul, known as the Earth Consciousness, resides inside the huge globe of Earth. Similarly, there is a great soul, a great spirit that exists within the sun. This means that inside the sun that pours physical light down on us there dwells a Spiritual Sun that emits spiritual Light.

The Spiritual Sun that shines down on the other world is in fact the spiritual body of the sun that pours its light down on the Earth. This means that the sun does not merely supply this world with light and heat, but also provides the other world with genuine spiritual energy. The Earth is a member of the solar system and is allowed to exist through the energy it receives from the great consciousness that rules the system.

The energy body that resides within the sun is the Stellar Consciousness of the solar system in which we live. It exists in the eleventh dimension and is known as the Solar System Consciousness. This Solar System Consciousness transmits light of seven colors to Earth through the three Planetary Consciousnesses of the tenth dimension. These are the Grand Sun Consciousness, the Moon Consciousness, and the Earth Consciousness. This energy is then passed through the ten spirits of the ninth dimension where it is separated into its spectral components and its flow directed to this and all the other worlds.

However, the inhabitants of the fifth dimension do not know all these facts. All they know is that the Spiritual Sun provides them with life energy as much as the physical sun does on Earth. They know that the light energy that they receive from the Spiritual Sun sustains their life and that they are able to live, thanks to this energy. As a result, they never forget to demonstrate their gratitude toward the Spiritual Sun, and even before they develop a simple faith in the Divine, they are extremely grateful to the Sun. They can often be seen in the early morning and evening, praying to the Sun with their hands together. In this way, the fifth dimension is the place where the flow of light can be perceived.

### 5. Love

Another characteristic of the fifth dimension is the awakening of a new level of the feeling of love. Here on Earth, we experience the love between man and woman, the love between parent and child, the love between friends, and the love between master and disciple. But the love experienced by the inhabitants of the fifth dimension is much purer.

Love is rather difficult to express on Earth, but in the Real World it is much more substantial. When one person loves another, a wave of emotion is transmitted directly to the recipient. The person who is being loved can feel it very strongly and this results in them experiencing the joy of the soul.

On Earth, love cannot be felt so clearly, one can never be sure if one is loved or not; this gives rise to a lot of romantic suffering. The source of problems of relationships between the sexes often lies in this sort of emotional tightrope we are forced to cross. You do not know if you are loved or not, and even though your partner may really love you, you might simply put it down to a desire to be nice to everybody or even feel that they do not hold any special emotion toward you. In the fifth dimension, there is a clear indicator or barometer that shows whether or not you are being loved and as a result your partner's emotion can be felt directly. It is like the difference between turning on a fluorescent light or a regular incandescent light bulb, or the difference between a sixty-watt, one-hundred-watt, or two-hundred-watt light bulb. In other words, the size of your partner's love is transmitted directly to you.

In this way, the Realm of the Good of the fifth dimension is a place in which thoughts are transmitted instantaneously and precisely—for this very reason the inhabitants of hell are not tolerated there. The spirits of hell are filled with hatred, envy, complaints, rage, and insatiable desire; if these emotions were to be broadcast to those around them, the place would soon cease to be considered heaven. The people of the Realm of the Good in the fifth dimension all share the emotion of love. Be it to a greater or smaller amount, of a higher or lower variety, love is something that is common to them all. In other words, each and every one of the inhabitants is like a power source that transmits the current of love to those around them.

Sometimes, the superior beings of the sixth dimension instruct the people of the Realm of the Good in the fifth dimension in how love can be grasped as an actual sensation and they explain exactly what it means. They are taught that when you are loved by another person, love will flow into you like an electric current, warm your heart, and fill you with joy. This love is in fact a facet of Buddha's mind. Although the people living in the fifth dimension are still unable to have a clear understanding of Buddha, they are able to feel the truth on a basic level. When the high spirits teach them about Buddha they do so by saying: "You feel love, don't you? Well, the greatest love that exists is the Sun that floats in the sky of the Real World. It provides us with warmth and energy without receiving anything in return; it supplies us with the energy of life without asking for a single penny. That benevolence, that compassion, is the true nature of Buddha. When you love each other, the vibrations of love fill your heart and you become very happy. This proves that you are Buddha's children and that, essentially, you are part of Buddha's life form." This is the message that high spirits patiently try to teach the inhabitants of the fifth dimension.

This is the first step in teaching about love. It may not explain about giving love to others, as is done in the Bosatsu Realm, but it does instruct in the basics of what love is, what it is to be loved, and what it is to give love. Those living in the fifth dimension learn that it is better to be loved than not and what a wonderful thing it is to love others. Eventually they will realize that love cannot co-exist with a desire for self-preservation, with a feeling that one's own well-being is all that matters.

### 6. Sorrow and Pain

People have always said that heaven is an eternal paradise where pain and sorrow are unknown, but is this really the case? Is it true that pain and sorrow do not exist outside hell?

It is generally believed that pain and sorrow are unique to hell and have no place elsewhere in the Real World. People cry, but is this something that Buddha had planned? Do people in heaven do nothing but laugh? This is the point I would like us to think about now.

Nobody can deny that joy, anger, sorrow, and pleasure are the basic emotions of humankind. For instance, joy is the opposite of sorrow; but if we were to say that sorrow is merely the absence of joy, it would not provide us with a sufficient explanation. In the past there was a lot of debate between the philosophies of monism and dualism. The monist movement stated that "evil is the absence of good" or that "cold is the absence of heat" and one of the thinkers of New Thought, Ralph Waldo Emerson, was one of its main proponents. While this philosophy is true to a point, it does not explain everything.

Certainly cold is an absence of heat and evil an absence of good, but there is more to it than this. We all know that we cry when we are sad, but we cannot say that the phenomenon of tears necessarily represents the absence of joy. We do not cry merely because we are not happy, we cry because we are sad, so sadness must be an emotion that exists in and of itself.

Next we come to pleasure and its opposite, pain. Does pleasure alone exist in the Real World and not pain? In the absence of pleasure is there pain? This is a question we must now look at. Pain does exist. If you play tennis or some other sport, you may reach the threshold of pain, but afterwards you will feel refreshed and invigorated. You experience physical fatigue but this in itself leads to the feeling of exhilaration.

In this way, it must be said that, for the most part, this world and the next are of a dualistic nature. The Ultimate Buddha, or the Supreme Consciousness, is monistic in that He consists only of the good elements, light, goodness, and love, but it must be remembered that He made this Earth and the lower spiritual worlds of the fourth and fifth dimensions to promote the development and progress of the soul. A world of relativity, including that of pain and pleasure, helps promote the advancement of the soul, and without it improvement would be difficult. Although a monistic world that knows only happiness might seem attractive, in a way it would be a very drab place.

This is why Buddha found it expedient to allow Earth and lower spiritual realms to experience that which appears to be sad or painful. A spirit in the fifth dimension does feel the pain of not achieving self-realization. People in the fifth dimension pray in much the same way that we do, but they sometimes do not attain all they desire. It is difficult for them to know whether they are praying rightly or not, but from a perspective of the spirits from higher dimensions, these prayers are sometimes judged as premature, and therefore no answer is given. If prayers are not answered it creates sadness and pain, but it could be said that, in this way, the inhabitants of the fifth dimension are undergoing spiritual training to strengthen their souls.

## 7. Nourishment of the Soul

There is a school of thought that claims that pain and sorrow are only an illusion, that neither pain nor sorrow exist in fact and that it is all simply a delusion of the mind. This is something with which I do not quite agree.

The Primordial Buddha, by His very being, is perfect and lacking in nothing. He is natural, He is supreme love, He is supreme bliss. In other words, He is the ultimate good, the ultimate truth, the ultimate beauty. As long as the Primordial Buddha embodies these characteristics, He can know no progress, no development, and none of the happiness that is associated with these forms of change.

So can we think of the Primordial Buddha's creation of the universe as that of a gardener creating a landscape? He may place a rock here, dig a pond there and release fish into it, plant trees of varying heights and make some of them bear fruit, maybe even plant a few weeds. He tries in various ways and makes every effort to improve the scenery. What appears imperfect to humans may be something that the Primordial Buddha has done on purpose to create a particular ambience in the garden. He creates soaring peaks and deep wells, flowers and weeds, tall and short trees. All combine to produce His garden in the way He extracts the most pleasure from it.

Like the various features in a garden, pain and suffering are allowed to exist under certain conditions. As I said in the last section, sorrow does not necessarily mean an absence of joy; we do not start crying just because we are no longer happy. When we cry it is a result of the positive presence of sorrow. In the same way, we do not feel pain as soon as pleasure ends, and some suffering springs from an identifiable cause.

Pain and sorrow do exist, but it is not because they are supposed to be good things. The true reason that pain and sorrow exist in this world and in the fourth and fifth dimensions is because they allow us to make a great leap forward. When things do not turn out the way we like, we experience pain and sorrow. When our efforts result in something quite different to what we intended, tears of sorrow may come to our eyes. But the tears are not shed for their own sake; the tears are to lead us to a higher level. Therefore, we must not look upon this world as being a world of pain or of sorrow, rather we should recognize that these feelings exist to act as a whetstone upon which to polish our souls, and we should strive to produce something better. It is similar to the process in which a gemstone finally starts to emanate brilliant light after being cut and polished.

Sorrow and pain exist, but they do not last forever. They are only fleeting experiences that are allowed to exist to act as the nourishment of the soul. Eventually we all travel to a world of joy and pleasure, a world of everlasting happiness, a world of ever-lasting paradise. Therefore, you must realize that the suffering of this world is only allowed to exist as the nourishment of the soul.

## 8. Divine Beings

I have suggested that pain and sorrow act as nourishment for the soul, but what should we do when we are feeling buffeted by the tribulations of the world? I wonder if you have ever heard how, when we seem to be at the very depths of despair, and when things could get no worse, a single ray of light can shine through the darkness. We tend to think that light always shines from above, but when we strike through the base of our sorrow, light can pour in from below. Shakespeare wrote numerous tragedies, and one thing we see from them is the possibility of light at the bottom of every misfortune. Once we break through the base of tragedy we will find the truth of humanity. In the recess of this truth lies an inner light. In other words, Shakespeare knew that not only comedies with happy endings lead people to progress, and that although something may appear to be a tragedy, in many cases it may provide a shortcut to the light.

There are many people in this world who curse their fate, saying: "Why am I always the one to suffer? Why do I have such bad luck?" They may have lost their parents in childhood, they may not have been able to afford to go to school, they may not have married or, if they did marry, they may have been separated by death or by choice. Perhaps they could not have any children or their children died young. Maybe their children became delinquent. There is no end to the reasons why people suffer; the seeds of sorrow are without number.

Is that to say that sorrow and bad luck are totally meaningless, and of no use to us? The world where we were before we are born on Earth, called heaven, knows very little of this pain or sorrow. People cannot always achieve their ambitions, but evil as such does not exist in heaven. In comparison, on Earth we seem to live at the mercy of fate, and ill-fortune can descend upon us when we least expect it. However, we should not forget the story of Job that appears in the Old Testament. He is forced to experience so many disasters that in the end he curses God, but God replies: "Are you so wise that you can judge the will of God? Be more humble, do you really understand my intentions?" These are the words that God used, but His real message was, "God uses various props in order to achieve evolution." People who have died, leaving us behind, may now be living marvelous lives in the next world; this just goes to prove that we must not judge things from a solely worldly point of view. The people who are faced with the heaviest trials are those nearest to the light. In the middle of joy, heaven is near, but in the depths of sorrow, too, heaven is close at hand.

When we manage to break through the bottom of sorrow and grasp the light, then heaven is manifested before us. This is a fact that all the people of the world should know.

## 9. True Nobility

I would like now to discuss the nobility of the soul. When do we feel the nobility of the soul or nobility in a spiritual sense?

Let us say there is a man who was born the only son of a rich family living in a large mansion with a swimming pool in the garden and possessed of lots of servants. He is brought up wanting for nothing, he is clever, good-looking, and is loved by women. Upon leaving school he is well-liked by all, and succeeds at everything he turns his hand to. However, does a person like this impress the people he meets with the nobility of his soul? Looking at people like this, do we feel that they have noble souls? Do the circumstances to which a person is born make him great? I do not think so. I feel that one of the origins of greatness is experience gained through adversity. It is only when a person succeeds against all odds that he becomes recognized as great.

For instance, there is Dr. Albert Schweitzer, an Angel of Light who struggled against the deprivations of the African jungle to heal and to spread the word of God. There is Thomas Edison, who, despite having only three months of formal schooling, went on to become one of the greatest inventors the world has known. There is Abraham Lincoln, who was born to a poor family but through his own perseverance and hard work went on to become president of the United States. Mohandas K. Gandhi, the father of India, was another Angel of Light, who took on the might of the British Empire and succeeded in winning independence for his country.

When we look at the lives of any of these people we can learn the real meaning of suffering or hardship. These lives show us that suffering and hardship are not obstacles, but serve to make the course and orbit of our lives more beautiful.

Of course, this is not limited to contemporary people, and examples may be found throughout history. In the case of Shakyamuni, the founder of Buddhism, he was born to earthly nobility, as a prince, and he wanted for nothing. Then at the age of twenty-nine, he threw it all away and set out on the road to enlightenment. In this way, when we set out to achieve a greater purpose in defiance of all difficulties, the nobility of the soul shines out, and this nobility becomes a light that will shine on future generations.

It is a great comfort, as there is no telling when we may ourselves be faced with suffering and hardship. The past is filled with people who were confronted with difficulties. Those who were overpowered by them were soon forgotten, whereas those who were able to repel all hardships without defeat achieved a nobility of soul that shone like a medal. If the life of Jesus had consisted of nothing but hardship, his name would not have been remembered, but he was able to triumph over hardship and manifest the nobility of his soul, and in so doing he became a beacon to guide future generations.

## 10. A Time for Guidance

If a person lives in a wonderful environment, they should be grateful and feel obliged to make the effort to improve themselves even further. If they are better off or more gifted than others, then they should make that much more effort to polish themselves.

If, on the other hand, a person lives in a worse environment than others, lacking in ability and wealth, if, whatever way you look at it, they are at a disadvantage, either handicapped or suffering from a congenital illness, is it only unfairness, or is it that such trials are ways to evolve and develop our souls? Even if we were to complain or grieve for our misfortune, how would this help our development? It is only when you advance bearing your cross that it will nourish your soul. It is only through living that kind of life that you will begin to shine.

Of course, this is not to say that you should go out of your way to pursue trials and tribulations. There is no need for you to pray to God to inflict you with hardships, but you should strive and cultivate a tenacity of spirit that will permit you to overcome any hardships should they befall you. There is nothing to be gained through complaining about what you do not possess. You should discover the splendor of what you have been given and use it as a weapon to fight off misfortune. There are those who are blind but who can speak most eloquently, those who cannot walk but whose hands are nimble. Some people are not so clever but are very fit while others are very intelligent but cursed with illness.

Before comparing yourself to others, complaining about what you feel you lack and grudging them their good fortune, you should look to what you have been blessed with and try your hardest to develop it. If you do this, you will find some clue to help you overcome your own problems. Each of our lives are filled with mysteries and problems, but somewhere there is always a clue to help us on our way. If you were to look at yourself through another person's eyes, I am sure that you would find that in some aspect you are superior to others while in another you are inferior. If you compare your character, talents, or physical features to others, you are sure to find some areas in which you excel and others in which you lack. This cannot be helped, but you should look upon your abilities, or lack of them, as clues to help you face the problems this life offers.

It is important to ask yourself why you should be confronted by your problems, as therein lies one of the objectives of your spiritual training in this life. There are all kinds of handicaps, physical and mental, but whatever they may be, they make clear the purpose of your life and your mission in this incarnation. When you realize this, you have been "guided." When you awaken to your fate, and determine to fight against it, you will become filled with courage and strength. At this time, the various high spirits of the other world, including your own guiding and guardian spirits will provide you with great strength. Therefore, the first thing to do is to search for the clues that are hidden in whatever life brings you. When you use these clues to solve your problems you will be given the strength of the high spirits.

When we become aware of the true essence of the soul, we often realize that it is our task to strive through eternity. Yes, humans live for eternity and it is only by mastering problems set before us that we can achieve the nobility of spirit that leads to Light, and to the nourishment of the soul.

Becoming aware of the nature of your own problem, and finding clues for solution—treasure this moment of guidance.

# Three: The World of the Sixth Dimension

## 1. The Main Road to Evolution

In this chapter I will concentrate on life in the Realm of Light in the sixth dimension. In the last two chapters I talked about the fourth and fifth dimensions where people usually go shortly after death. Now I would like to move on to the sixth dimension. It is known as the realm of the high spirits, because since ancient times it has been said that this is home of the gods. But what kind of beings are these gods? I would like to start with a brief explanation of these beings.

Of course, although I call them gods, I am not referring to the Creator who produced the universe and all in it. That being is not to be found in the Realm of Light in the sixth dimension. The gods I am now referring to were born on Earth and took human form. They are those who were so virtuous and the work they carried out so great during their time here that other people did not consider them to be ordinary human beings. People could feel that they were different and guessed that they must be close to

God. One example of this kind of person is the famous Japanese scholar and politician Sugawara-no-Michizane (845–903) who was enshrined as the god of learning soon after he left this world. On the whole, people who have contributed to society while on Earth and whose hearts are not attuned to hell are allowed to enter the sixth dimension.

Therefore, it can be said that the people allowed to enter the sixth dimension are those who earned the respect of others while living on Earth, but what kind of people receive respect from others? They are very virtuous and capable of achieving goals that appear impossible for the ordinary person to manage. In other words, they are more evolved than other people. As a result, they are so spiritually powerful or so capable of achieving great things that they strike awe in the hearts of others and are revered as gods. Among their numbers may be found many of the Japanese gods who make up the pantheon of the Shinto religion.

## 2. Knowing God

I have already stated that there are many people living in the Realm of Light of the sixth dimension who are known as gods, but what exactly do we mean by the word "god"? This is a question that has long been debated by men of philosophy, religion, and theology, leading some to declare that, "to know God is to know everything," while others counter this, saying: "man was not made in the image of God, rather God is a figment of the imagination created in the image of man." However, none of these wise men were ever able to offer a definitive meaning of the word "god." I would like, now, to offer my own current idea as to what god is. First, I would like to divide the beings that go under the title of gods into two categories; on one hand we have God the Creator and on the other, all the other gods.

Christians believe in a Trinity of the Father, the Son, and the Holy Spirit; each of these three "persons" may be considered to signify a facet of God, whereas in some cases only the "Father" is identified as God. In general, people consider a god to be an existence that is spiritually superior to human beings and by this definition of the word it is true to describe the Holy Spirit as a god. In other words, the inhabitants of the sixth dimension are what we know as holy spirits whereas God the Creator exists on an immeasurably higher level.

One must not assume, however, that all the inhabitants of the sixth dimension are gods simply because they dwell in the Realm of Light. The people in this world find it difficult to understand, but the sixth dimension is divided into multiple levels. Of course, this does not mean there are transparent barriers in the sky, each supporting its own population like the floors in a terrestrial highrise apartment building. It must be remembered that the inhabitants of the sixth dimension do not have corporeal bodies as we do, but are living consciousnesses. A consciousness is a form of energy that can be likened to electromagnetic waves, electrical energy, or a gaseous form. It is an energy form with specific characteristics and individuality—this is the true form of the spirits in the Real World. Therefore, when I say that there are different levels within the sixth dimension, it means that the inhabitants have different wavelengths. They are not divided simply in a vertical order, but the divisions of the upper and lower stages of the sixth dimension occur due to the different wavelengths of the inhabitants.

To make this easier to understand, let us take the example of filling a cup with muddy water. If you waited, you would see the heavier particles drop to the bottom while the water at the top would gradually become clear. The nearer the top of the glass you looked the clearer the water would appear, while the lower you looked the darker it would appear. In the same way, the cruder wavelengths, that is to say the beings that carry the weights of materialism of this world on Earth, sink to the bottom whereas the less attached, purer spirits who are nearer to godhood, rise to the top. Whether you think of consciousnesses as wavelengths or energy forms, you should remember that they all exist in different places according to their individual properties.

### 3. The Stages of Enlightenment

The first step toward enlightenment is to learn that the inhabitants of the Real World are living consciousnesses who are divided into different levels according to their degree of development.

The word "enlightenment" has many meanings. One of the basic meanings of enlightenment is to grasp an understanding of the fact that "people are more than their bodies." Simple as it may seem, this is in itself a form of enlightenment. A large number of the inhabitants in the Posthumous Realm of the fourth dimension have still not understood the fact that a person is not just a body, and they live their lives in a semi-corporeal, semi-spiritual way, unsure of their true selves.

Another form of enlightenment is when a spirit rises from hell to heaven. This enlightenment is a minimal understanding that there is more to life than self-preservation and that we should live for others. The inhabitants of hell live in selfishness, they think only of themselves and do not care what happens to others as long as they are all right. They cannot understand what is wrong in living only for their own sakes. However, after a few decades or centuries of living in hell surrounded by like-minded people, they gradually become disgusted with their rapaciousness and the time comes for them to make a change. This is the first enlightenment that is necessary to make the transition from hell to heaven. The time comes when they repent and long for a world of peace and quiet. When this happens, they return to the Astral Realm in the upper part of the fourth dimension and then eventually they graduate to the Realm of the Good in the fifth dimension, as I explained earlier.

The fifth dimension is known as the "Realm of the Good" or the "Spiritual Realm" and people who have awakened to their essence as souls, to their spirituality, to the importance of good, live there. However, although they may be good people, they are still not serious about attaining enlightenment. Their hearts and minds have still not turned fully toward the Divine.

In contrast, no sign can be seen of atheists in the sixth dimension. All the inhabitants realize, to a greater or lesser extent, that there exists some kind of great being that gives life to people and spirits. Depending on their understanding, different individuals may call this existence Buddha or God and they all use different methods to advance toward the Great Being.

Many priests of all the major religions can be found living in the sixth dimension where they struggle to answer questions of divinity. There are also inhabitants whose main interest is not necessarily religious matters. These are people who managed to evolve through their professions while living on Earth. They had not wholeheartedly pursued God or Buddha, but had attained such a high level of development through their work that they were able to enter the sixth dimension.

A great number of scholars reside in this realm. There are many university professors and other first-rate teachers whose minds are not attuned to hell. There are many doctors, lawyers, judges and such like, from respected professions on Earth, as well as politicians, bureaucrats, and other people with pure hearts. When you look at them, it is obvious that they are highly evolved in some particular field. For instance, a large number of artists and musicians live there, having developed their artistic talents while on Earth.

People's professions, while living in the sixth dimension, are related to the ones they were involved in while on Earth. Those who used to be leaders of various religions offer religious direction to people on Earth, ex-politicians offer advice to terrestrial politicians. Ex-bureaucrats help people on Earth working in government offices, those who were artists send inspiration to artists, while those who used to be in teaching positions guide students on Earth. In this way, they continue working in their own specialty while striving for enlightenment. By passing on what they have attained, they are able to experience an early form of the work of the Bosatsu in the seventh dimension. In other words, they are able to experience what it is like to work for the sake of others.

To sum up, the most valued quality in the sixth dimension is that of being useful to others. To be of use in this world, to help further its development and progress, this is what the people live for. While this quality is still at too early a stage to be described as true love, it represents the preceding step and leads to the germination of true love.

### 4. The Ocean of Light

I would like now to give a description of the sights in the sixth dimension to help you visualize it. As I have stated in numerous books, the higher one travels through the dimensions, the brighter it becomes, so when one arrives in the Realm of Light in the sixth dimension, it is quite dazzling.

I am sure that you must have walked or driven through the mountains and suddenly come to a break in the hills that presented you with a glimpse of a town down in the valley or a distant view of the sea. The sudden change of scenery takes your breath away and this sensation is similar to the one people experience when they look upon the sixth dimension for the first time. It resembles an ocean of light. At first it is too bright to look at; it is like the sea that reflects the strong sunshine of the summer, and it takes some time for people to accustom their eyes to it. I am not merely talking metaphorically when I say this, as there actually is an extremely beautiful ocean in the Realm of Light in the sixth dimension. I stated earlier that the sixth dimension is divided into various levels in the vertical plane, but there are also divisions in the horizontal plane, and it is this horizontal geography that I would like to explain next.

First there is the major heaven. The greater part of the sixth dimension belongs to the Realm of Light whose inhabitants are high spirits who have developed their souls in the right way. However, there are several other realms that exist outside the major heaven in the sixth dimension in an area known as the minor heaven, for instance, the Ryugu (dragon palace) Realm. This realm has been alluded to in Japanese legends where it is described as lying at the bottom of the sea and being the home of the gods of the sea. Many high spirits live there, and although it basically belongs to the sixth dimension, the Ryugu Realm also extends vertically down through the fifth and fourth dimensions. It is a vast world set within a sea of light, giving the impression that it exists under the sea. Geographically, it corresponds to Lake Biwa, Miho-no-Matsubara, and Matsue in Japan, all coastal areas of great beauty.

All kinds of creatures can be found in Ryugu. As it says in the old legends, not only people but dragons and other beings live there. Foremost among these are the Dragon Gods who serve the high spirits as emissaries. They have power over various natural phenomena, generating tremendous amounts of energy at major turning points in history. They are not human spirits, and they possess vast spiritual power.

Other realms in the minor heaven of the sixth dimension include the Sennin (hermit sorcerers) Realm and the Tengu (goblins) Realm. Whereas the Ryugu Realm is an ocean world, both the Sennin and Tengu Realms are for the most part very mountainous. The landscape is very rugged and many spirits may be seen there engaged in ascetic discipline. When these spirits lived on Earth they had actually struggled to achieve enlightenment, but they had concentrated on physical asceticism in an effort to attain supernatural powers. In other words, the inhabitants of these realms achieved enlightenment solely through the use of metaphysical forces and tend to lack in human warmth and kindness.

## 5. Eternal Travelers

So far I have talked about the sixth dimension focusing on the theme of evolution. I am sure that there are some among you who realized that, basically, humans are eternal travelers. Others may disagree, and ask why they have to exert themselves to evolve all the time, or question what is wrong with the way they are now. I suppose, in a rather shortsighted way, they have a point.

However, when considered from a spiritual point of view it can be seen they are wrong. The reason we have to develop is because human life is infinite. Our true lives last for tens of thousands or tens of millions of years and, if there was no change during this long life, our souls would stagnate. Having stagnated, we would know no further joy. In other words, we would become bored. It may be fine to do the same thing for one or two hundred years, but not for eternity.

As long as we have a personality and consciousness, we feel we have to achieve something. Take the average office worker, many of them spend their lives dreaming of leaving their company and becoming free, but many of them remain where they are until retirement and then, when they are finally free to do whatever they like, they find that time hangs heavily on their hands. They may be free, but they have nothing to do. This gradually becomes unbearable, and the majority of people last less than a year before looking for another job or immersing themselves in their hobbies. This is due to the essential nature of our souls that are naturally industrious and are not capable of sustained sloth. That is why, although we may be tempted to take time off, we cannot remain inactive for long periods. People are made to work.

In other words, the soul is industrious and creative in nature. Many people claim that they would be happier if they did not have to work, but if they have their jobs taken away, they are at a loss as to what to do. The soul is a diligent and industrious worker so it is only natural that people should strive toward advancement. Nobody is content with doing a half-hearted job. In order for the soul to be satisfied and know peace and happiness, we should constantly strive to do a more complete job. In this respect, it can be said that the true state of the soul, or a person for that matter, is that of an eternal traveler.

## 6. Uncut Diamonds

In this section I would like to elaborate on the theme of the eternal traveler. Granted that humankind is a race of eternal travelers, devoted to the development and evolution of the soul, but why should there be difference of rank between the high spirits and the lower spirits? This is a question that many people have. Why should there be superior and inferior people? Why should there be Angels of Light and ordinary spirits? Buddha loves all people equally so surely it is strange that there should be these differences in rank. This question lies at the bottom of many people's minds and my answer to it is the title of this section, "Uncut Diamonds." People are all created like diamonds that will shine if they are polished. When these diamonds are first dug out of the mountain they are still in their rough state and it is left up to the individual to decide how they want to shine. This is a task that nobody can escape.

People often compare the Guiding Spirits of Light with ordinary spirits, saying that the Guiding Spirits of Light might shine like diamonds but ordinary people are like pieces of coal or charcoal, or stones that will never sparkle in the same way. While the stones on the riverbank are very different from diamonds, it is not accurate to describe the nature of the soul in this way. The proof is that no matter what kind of person one may be, if the soul is polished, it will shine.

When people on Earth hear stories about the spirits of hell or the devil, they often wonder why such beings are allowed to exist at all. Many think that Buddha should drive them out of this world, out of hell, and banish them to some distant corner of the galaxy. However, they feel like this simply because they do not know the true nature of the soul. They think that the creatures of hell are ugly in the extreme and work endlessly to bring about evil, but, given the chance, even they may be saved. I have met many people who have been possessed by bad spirits, and I have spoken directly to the spirit that has them under its sway. Each time, I was reminded how little these spirits know of the Truth. For instance, they do not know they are spirits, nor do they realize that the human body is not their true form. They do not know that they are supposed to do good or that they are presently living in hell. The spirits of hell are simply ignorant of their situation.

However, if they are taught the Truth they sometimes return to themselves. They realize that they have been living a false life and that they cannot go on in that way. They make the decision to live a proper life. At that moment their spiritual body, which up until that time had appeared totally soiled, suddenly begins to shine and a halo forms around their head.

Why should their bodies start to shine in this way? If these people were really no different from the stones on the riverbank, then they would not shine no matter how hard they were polished. However, the fact that they do shine shows that even malicious spirits, even Satan himself, are diamonds at heart. They are uncut diamonds and that is why they are able to shine if polished. These diamonds were merely covered over with impurities. The average person would want to throw them away, thinking that they were just ordinary rocks. However, if they are washed in the river and properly cut, they will shine brilliantly. This is an example of the infinite potential that is given to us by Buddha, and which is a manifestation of His boundless love for us.

### 7. The Essence of Politics

Many of the people living on Earth are filled with the ambition to become great men and women in a worldly sense. They may think that politicians are most powerful and strive to join their ranks, perhaps even rising to become ministers or reaching the pinnacle of power as prime minister or president. However, while we share a desire to be looked up to, to become a member of the ruling class, at the same time we tend to ridicule politicians and politics. We feel a certain disgust at the naked ambition they personify and look upon their behavior with contempt.

At present, people have lost sight of the true meaning of politics. Politics is basically a form of hierarchy, the relationship of the ruler and the ruled, of authority and compliance. It is basically a form of power pyramid with a few chosen people standing at the top who are supported by a large number of people at the bottom. Being triangular in shape means that it is a very stable structure. If it was circular, it would soon tumble to one side or the other.

This form of organization is not restricted to politics, and the vast majority of corporations also incorporate the pyramid structure, with a large number of office workers at the base. As we move up through the echelons of management to the top, there is, invariably, a single president. Even schools have the same structure, with many teachers at the bottom and the principal at the top. This type of pyramidal structure can be found in every conceivable field of human endeavor, but why should this be? The answer is that it is a reflection of the structure of the spirit world. There are more people in the fourth dimension than in the fifth, more in the fifth than in the sixth, and so on until one arrives at the ninth dimension where only ten spirits abide.

Human beings live a communal life and this demands the existence of leaders who are able to maintain unity. If everybody just spoke his own mind there would be no coordination, and it would be impossible for a community to act as one. In other words, politics as we know it springs from a need to supply leaders, and the spirits from the Realm of Light of the sixth dimension are striving to improve their qualities as leaders. They have been engaged in a variety of professions on Earth but what they all have in common is the ability to lead. They are all people who are looked up to for their proficiency in guiding others.

## 8. Unquestionable Power

Having said that it is the people of the sixth dimension who become the leaders, I would like now to think about why it should be that they are able to do this. Why should one person have the power to make others bend to his will? What gives them the right to control or teach others? What is the origin of political or spiritual power? The truth is that spiritual power is something that is bestowed upon people from above; in other words, it is the power that originates in Buddha.

If Buddha were on the side of the denizens of hell, there would be truth in what they say; but obviously He is not. Looking at which side He supports, people are able to judge which direction is correct, which opinion is true. In other words, Buddha's presence is like the North Star, immovable in the sky, that shows people the direction in which they should move. People who stand close to Buddha are able to command the loyalty of those below them and take on a leadership role. This means that the power of the numerous leaders who come from the sixth dimension originates from the power and wisdom of Buddha. Without Buddha's power, nobody can have authority to lead others either here or in the Real World. It is only once people have won Buddha's approval of their ideas that they can feel courage and brilliance.

The people of the sixth dimension are aware they have been chosen by Buddha and that they are members of an elite, so they feel an obligation to help those less developed than themselves and offer them guidance. There is a tremendous variety of ways to teach others, and choosing that which is best suited to them, the spirits of the sixth dimension study the will and intention of Buddha in order to pass this knowledge on to others. In other words, the sixth dimension is a place where spirits study and explore the will of Buddha as their main task.

As a result of this study, they receive the power to guide others. They are able to say, "From what I have learned of Buddha's will, we should adopt the following policies." "We should consider these economic principles." "We should create the following type of art." "We should carry out the following educational reforms." They are able to speak with perfect confidence because the origin of all power lies in a knowledge of Buddha's intentions. The very basis of power lies in a knowledge of Buddha's mind.

To understand Buddha's thoughts, to gain "knowledge," is the most important objective of the sixth dimension. This "knowledge" represents an understanding of the Truth of Buddha and is a vital element of existence in the sixth dimension. One is unable to remain in the sixth dimension without acquiring a knowledge of the Truth, and the primary requirement for admittance to the sixth dimension is an eagerness to acquire knowledge of the Truth.

## 9. Inspiring Words

I have been talking about the source of unquestionable power, and now I would like to think about words. There is a famous quote from the Bible that reads: "In the beginning was the Word, and the Word was with God, and the Word was God" (John 1:1) This quote serves to illustrate the importance of words. When the Guiding Spirits of Light come down to Earth, it is through words that they manage to persuade and move people. Of course, they sometimes resort to miracles but these supernatural phenomena alone are not sufficient to instruct the people. While miracles may be useful expedients to attract people's attention, they do not in themselves lead others to true enlightenment.

I would like to ponder now on the reason why some words have the power to move us as they do. When the Guiding Spirits of Light preach or, to use the modern vernacular, give lectures, why should it be that they have the power to move people's hearts? Why do they cause tears to come to the eyes of their audience? This is something that we all need to understand.

The reason we sometimes cry when we hear a sermon in this world is that deep in our hearts exist memories of other occasions when we listened to the Truth in previous lives, and this creates a deep emotion within us. It may be memories of listening to Shakyamuni preaching in India or of Jesus in Israel. There have been countless previous times when we have heard the Truth preached on Earth, not to mention the times that we heard the Guiding Spirits of Light teaching it to us in the Real World, that is to say in the fourth dimension or above, and it is these memories that make us want to cry.

People have an instinctive understanding of what is good for them. We do not cry simply when we are sad, we also shed tears of joy or emotion. The tears that we cry upon learning of the Law or attaining enlightenment are known as "the rain of Dharma"; they have the power to cleanse our hearts and purify us through removing worldly desires. As the rain from the sky washes the dust out of the air, purifies the atmosphere, cleans the earth, and drives away the fog, so "the rain of Dharma" washes away our sins as it runs down our cheeks and allows our hearts to shine—to shine with the radiance of diamonds.

Religious leaders on Earth should provide many opportunities to bring forth "the rain of Dharma." Conveying powerful words in writing is important, but even more important is giving speeches that touch a string in the hearts of people who are listening, and move them to tears. When this happens, people experience something that is not quite of this world, they feel a strong aspiration to the path of enlightenment. In other words, they recall the fact that they had been greatly moved by talk of the Law, of the Truth, in the past.

Words can be described as a method of expressing the writer's or the speaker's degree of enlightenment. The deeper this enlightenment, the stronger the influence they have and the greater their power to move the hearts of the readers or listeners. The words of an unenlightened person have no power to move; the writings of an unenlightened person leave the reader cold. But when an enlightened person writes on the same subject it warms our hearts and fills us with excitement. This is because their enlightenment reveals itself through their writing.

If you ever feel the need to test your own level of enlightenment, you can do so by seeing if you can speak powerfully on the subject of the Truth. The higher your level of enlightenment, the more powerful your words will become. Light will spring from them and they will move the hearts of your listeners. This is a useful way of checking the progress you have made in your spiritual training.

## 10. And So to the World of Love

I have tried to present you with a rough outline of the sixth dimension. Vertically, it is divided into upper, middle and lower areas but horizontally we find the Realm of Light in the front, the Ryugu Realm in the middle and the Tengu and Sennin realms, inhabited by those who devote themselves to physical discipline, in the rear. People who like to boast of their power in this world go to the Tengu Realm, while the Sennin Realm contains spirits who specialized in perfecting supernatural powers.

As I mentioned earlier, the sixth dimension contains a large number of spirits who are thought of as gods. In fact, the vast majority of beings who are worshipped as gods belong to the upper part of the Realm of Light in the sixth dimension. They are known as "special-mission high spirits." Among them are the Buddhist deities known as the "guardian gods," and the many gods of the Shinto religion, such as the god of wealth. The enlightenment of these spirits is not necessarily limited to the sixth dimension. Some of the spirits have attained the enlightenment of the Nyorai (eighth dimension) or Bosatsu (seventh dimension) but choose to be in the sixth dimension in order to carry out a particular mission. This means that among the inhabitants of the Realm of Light are many Nyorai or Bosatsu who have been sent to govern the sixth dimension while supervising the guidance of the spirits on Earth. A large number of these spirits are so evolved that they truly deserve to be called gods.

Another group of spirits who undergo training in the upper regions of the sixth dimension are known as the Arakan. The Arakan are studying to attain the state of Bosatsu, working to clear the stains from their minds. Having corrected their mistakes through self-examination, halos form behind their heads. They are still at the first stage of the path to enlightenment, or on a preparatory level to the world of Bosatsu, and live in the upper part of the Realm of Light. They are called Arakan in Buddhism, but some ministers and priests of the Christian faith are also to be found on this level.

"Arakan" means those who have completed the curriculum of personal discipline. They are currently working hard with the aim of advancing further, to enter the seventh dimensional Bosatsu Realm. Arakan are struggling to master the Truth while at the same time striving to find ways in which to teach it to others. After they have virtually completed the required process of polishing their souls, they advance to the next stage—living for the sake of helping others, with love and compassion. This is how Arakan become Bosatsu; in Christianity, this is described as becoming an angel. These are the kinds of people that one can find in the upper reaches of the sixth dimension.

This is the main path to the discipline of the soul, and it can be said that one of the objectives of discipline in the Realm of Light is to attain the state of Arakan in preparation for becoming a Bosatsu. It can be seen from this that, without completing the disciplines in the sixth dimension, it is impossible to advance to the seventh dimension.

The same can be said of the spiritual training of people living on Earth. First we must gain a knowledge of the Truth and put it into practice, then we are allowed to live a life solely devoted to love. People must realize that there is no becoming a Bosatsu without absorbing the Truth. The first step is to absorb knowledge of the Truth, then you can aim to help other people through universal love, through the power of enlightenment. It can be said that this is the right path to follow, considering the way it is in the Real World.

# Four: The World of the Seventh Dimension

## 1. Love Overflows

In this chapter I would like to talk about the seventh dimension of the spiritual realm. Another name for the seventh dimension is the Bosatsu Realm. Obviously, this is a Buddhist name and is not used universally, but it is familiar term to Japanese people and so I will continue to use the term in this book.

The seventh dimension can be summed up as being a world of love. The word love is used every day in a variety of meanings and appears in all kinds of literature. It is one of the most basic desires of humankind and something that we go to almost any lengths to attain. Everyone wants to be loved by others. And it is the disparity between the desire to be loved and the feeling of actually being loved that decides whether or not you are happy. The subject of love has been discussed countless times in novels, poems, and philosophy as well as being a major theme in music and painting. However, there is nobody in all of history who has been able to provide a conclusive definition of love and, for that reason, in this chapter I would like to concentrate on the problems of love.

In the last chapter I explained that "knowledge" was very important in the sixth dimension and, moreover, that this "knowledge" was not merely an accumulation of Earthly knowledge, but a knowledge of the Truth of Buddha. What I now want you to understand is that one stage above the world of knowledge lies the world of love. It is said that love surpasses knowledge, but this is not to say that knowledge is unnecessary if you have love. Rather, it means that while knowledge is important, love is even more important. This is a wisdom that is gained through experience.

There are many people who, although they are kind and always willing to do things for others, fail to find happiness. This is because they suffer from what can only be described as "charity obsession." They try to help others for the best of motives, but find this only creates resentment and they become very unhappy. They devote their lives to others but receive no thanks for their efforts which leaves them feeling disappointed and futile. There are a lot of people like this and they are all victims of "charity obsession."

Love appears so simple, so easy, but in practice it can be very difficult because one of the functions of love is to guide and nurture people to develop. In order to do this, it is necessary to develop a deep understanding of people, of the world, of the true nature of people's hearts and minds, and of the will of Buddha. Therefore, we can see that love backed by knowledge can help sustain, nurture, and develop everything in this world, while love without knowledge is a fragile, delicate thing that soon withers and dies.

If we study the true essence of the human mind, we realize that love wells up like a fountain, it flows out from the innermost parts of the mind and we can use this understanding to provide us with a new starting point.

## 2. The Functions of Love

In this section, I would like to look at the functions of love. What are these functions? What role does love play in our lives? What part does love play? What would happen if love were to disappear? Is it something that always existed or did humankind feel a need and invent love to fill it? These are just some of the questions I would like to try and answer.

First, I would like to think about the energy that supports us through our lives, from the moment we are born until the day we die sixty or seventy years later. Let us look back over the time from when we were babies crawling around on the floor, through kindergarten, primary school, and on to high school. To begin with babies, their main task seems to be to reach out for their mothers' love. This appears to be instinctive, and they are aware of love from the very beginning. When a baby feels unloved, it begins to cry; but when it feels loved, it looks extremely happy. If a baby is given milk or toys, it is happy; but when its parent goes away or it cannot have its own way, it begins to cry. Looked at this way, we realize that although a baby's heart is pure and unblemished, it instinctively exhibits one of the functions of love. Although it is still immature, it is quick to recognize whether it is being loved or not.

At around the age of three, four, five, or six, a child begins to compare the amount of love it receives from its parents to that given to its brothers and sisters. Even though the child is still very young, if it feels that its parent's love is directed to a newborn brother or sister, it will react to this knowledge by misbehaving. This is the origin of jealousy and, as can be seen, it is an emotion that begins from a very early age. Looking at the way in which jealousy functions, we see that it stems from a powerful desire to be loved. When this desire is not satisfied, the child reacts by playing up and causing trouble for others. Love seems to act as a form of nourishment to young children.

As the child moves up out of primary school and into high school, it begins to want love from its friends and teachers as well as from its parents. When a child is praised by teachers for doing well in its studies and wins the admiration of its friends, it feels extremely satisfied. If a child is not very good at its studies but excels in sports, it will still be able to win people's attention and feel loved. A child who does well in sports or study will receive the attention of and be loved by the members of the opposite sex. It could be said that until we reach adulthood we rely on the attention of others to provide us with a form of nourishment.

So what happens after we reach adulthood? From our mid to late twenties, both men and women think seriously about getting married. In order to attract the opposite sex, we study hard at university or work hard at a career. A woman might wear beautiful clothes and makeup, searching for ways to make herself appear more attractive. All this illustrates our need to win love.

Looking at these developments, it would appear that the urge to be loved by others is something that we all desire instinctively. But we should question whether we should merely strive for receiving the love of others as we do naturally from the moment we are born.

### 3. The Dynamics of Love

This appetite for love I discussed in the last section is an example of the dynamics of love, of the interrelationship of love, or what could even be described as the law of cause and effect of love. I think that we need to consider the dynamics of love in greater detail.

Babies all thirst for love and their parents are there to provide it for them. There is a father's love and a mother's love; both differ and yet combine to provide a steady flow of love toward the child. As time passes, the child grows to adulthood and has children of its own upon which it in turn can focus its own love. Children bask in their parents' love and their very existence is a source of pleasure to their grandparents. The mere act of looking at their grandchildren's faces or holding their hands is sufficient to fill them with joy. Love is not only given, it is also in some way returned. In this way, it can be seen that love moves in a cycle that is completed every twenty to thirty years. Parents give all their love to their children, then these same children grow and have children of their own who they love in turn. Then the grandparents' love is given to their new grandchild.

So far, we have been looking at love within the family. But there is a much greater problem of love that faces us all—that is the love between men and women. From an age of about ten, people gradually become more interested in the opposite sex and from their mid to late teens they become filled with thoughts of the opposite sex. It is as if some strange force is pulling them together. Men and women find themselves attracted to each other and it becomes impossible for them to think of anything else. It is a very strange emotion. Although nobody tells them, a sense of obligation emerges between men and women. They feel that because they love each other, they should not become close with anybody else. It is very strange that, although there is no formal agreement, a kind of contractual relationship emerges between them. They seem to know instinctively that love will keep them together. From their teens to their twenties, men and women feel a love for each other that creates a kind of contractual relationship that, in turn, matures into matrimonial love.

Matrimonial love is an exclusive relationship that is protected by law and into which nobody can intrude. By observing a marital relationship, we can understand that exclusivity is part of the nature of love. If a husband or wife plays around and never comes home, it will cause their partner a lot of heartache. As you can see, it is true that there is a trace of a monopolistic element to love that excludes the intrusion of outsiders.

## 4. Eternal Love

I have said that matrimonial love includes an exclusive and monopolistic emotion; but does this possessiveness spring from self-preservation and does that make this kind of love wrong? There is a school of thought that holds that it is human nature to love all people equally, and therefore everybody should be treated the same. However, if a woman were to treat all the men she met in the same way she treated her husband, or a man were to treat other women as he did his wife, where would it lead? It would mean the destruction of their harmonious marital life.

A man and a woman live as a married couple based on a promise that ensures a long-term plan of building a home in which to raise their children. If people did not need to create a home, if the only reason that both man and woman exist were that of biological reproduction, and if it were the government's responsibility to raise children as in the ideal state Plato described, this would reduce the purpose of living as a couple to the mere continuation of the species. However, this is not Buddha's will. Buddha's perspective is that great benefits await the man and woman who work together for several decades to have children and create a home in which to raise them. Matrimonial love may appear to be exclusive and monopolistic but this appearance of selfishness and self-satisfaction is unavoidable; it represents the minimum evil that has to be committed in the name of domestic love and happiness.

Although this may seem a rather narrow kind of love, it is a necessary evil in order to create something of a higher dimension. Therefore, this monopolistic love that exists between a man and a woman does not have to be an evil thing. It is only when it is taken to extremes, when one of the partners loses their respect for the dignity and freedom of the other, when jealousy gets the upper hand, that it can lead to unhappiness. A certain amount of jealousy between husband and wife is permissible as long as it is kept within reason and is used constructively to hold the relationship together. But should one of the partners become excessively jealous and forever find fault with the other, it is certain to result in unhappiness.

So far, we have been thinking about the love between men and women. Through the experience of this love, we learn that the essence of Buddha is love. The way that men and women fall in love, marry, and experience matrimonial and then parental love is His way of teaching us about love. However, this love that springs up between men and women is not necessarily everlasting or eternal. To some extent it can be said to be instinctive; and, although the word "accidental" may not be quite appropriate in this context, there is no denying that some relationships do begin with a chance encounter that results in love blossoming.

Buddha has planned for men and women to marry, have children and create a home, but is that all there is to life? Of course not. The reason He causes us to love in this way is in order that we can awaken to true love. No matter how egotistical a person may be, they are sure to be well-disposed toward the opposite sex and to feel love toward their children. Feelings of love toward the opposite sex and one's own family are planted within us in order that we may learn of eternal love. The real reason for Buddha to bless us with these basic forms of love is to help us awaken to the real love of the higher dimensions.

## 5. For Whose Sake Do We Love?

I have already talked at some length about love. But the question I would like to pose now is, "for whose sake do we love?" From childhood, we realize instinctively that it is good to be loved and bad not to be. But if everybody is on the receiving end of love, there will be nobody left on the supply side. If there is only demand for love and no supply, then the world's pool of love will dry up. If everybody thinks only about receiving love, then the supply will run out and we will see a sudden rise in the number of people hungering for love.

Love does not only exist between men and women or in the home. When we go out into the world we find that love even springs up between strangers. Perhaps the word "love" is a bit misleading in this context; perhaps it would be better described as being well thought of by others. If you are thought well of by others it can be said that you are loved. In the same way, if you think well of people, look after them and are kind to them, you can be said to love them.

Looked at through spiritual eyes, the people of this world are like travelers trudging through a desert. They are all crying out for water as they stagger along under the burning sun. But if they could only give love to others they would find that their thirst is eased. As long as they think only of receiving love their thirst will only grow. Looked at this way, it is easy to see for whose sake we love. I am sure you have heard the expression, "one good turn deserves another," which means that if you do something for somebody else, you will benefit yourself eventually. Well, love is very much the same; if you love others, you will yourself be loved.

Have you ever paused to consider the economics of love? In many ways it can be said to resemble any other form of economics. For instance, let us look at a farmer. He grows vegetables and rice, which he takes to market and sells to get money to buy the things he needs. If he uses that money to buy a car, the man who made the car will get the money from the farmer, which he can then use to buy vegetables and rice. In this way, everything goes around in a circle. In basic economics, money acts as the medium through which a person's work is appraised and paid for, and this travels around in a ceaseless cycle. Love is the same; the love you offer passes through several others before returning to you. Therefore, the principle is that you will receive as much as you have given.

In the same way that rice or a person's job has a value accorded to it, we receive the same quantity of love as we offer to others. This may be impossible to see if your view is confined to this world on Earth. But looked at through spiritual eyes, it can be seen that people who love are in turn loved by others.

The more we love, the more love we can earn. That is why the higher spirits, the Guiding Spirits of Light, who love large numbers of people, receive a tremendous amount of love themselves. But if this is so, where does the love they receive come from? Is it the praise they receive from others? To some extent this is true, but the love they have given is rewarded by Buddha and returned to them through Him as His blessing.

## 6. The Essence of Salvation

In this section I would like to change the subject somewhat and think about the essence of salvation. I have already explained that the seventh dimension is the realm of love and that love takes various forms including domestic love, parental love, and the love that exists between men and women. Then, what form does the love of the inhabitants of the seventh dimension take?

Unlike the other types of love I have discussed, the love of the seventh dimension is not an instinctive love. When the Bosatsu of the seventh dimension are incarnated on Earth, their main task is to love others, regardless of whether they know them or not. They dedicate themselves to enlightening and saving the people of this world through the will of Buddha, and continue this work after they return to the Real World.

There are several schools of Buddhism that place a great deal of importance on the theme of salvation, relying on the power of Amida (Amitabha) to help practitioners save others and be saved. However, what exactly is meant by salvation in this context? I mentioned earlier that when looked at through spiritual eyes, this world resembles a desert with people staggering through the heat haze in search of water or an oasis. But if these are the people of this world, what is their salvation? Surely, however we think about it, the essence of salvation must be whatever quenches their thirst. But what is the water they crave?

Jesus Christ answered this question two thousand years ago. One day when he was thirsty, a woman offered him water. After he had drunk, he said: "You can quench your thirst with water, but the thirst will return; however, whosoever drinks from my words of life will never thirst again." His words were true and in them can be found the essence of salvation.

What Jesus refers to as the "words of life" are the truth that leads human souls to eternal life. In other words, they are the Truth of Buddha. People who live for the sake of the Truth of Buddha have eternal life; therefore, they do not become lost, they do not tire, they do not thirst.

Countless numbers of people have been blessed by the words of Buddha's Truth. It showed them the way, and they took courageous steps to advance toward Him. The essence of salvation lies in the Truth of Buddha that awakens people, the awakening that leads them to enlightenment. The essence of the work of the Bosatsu is to spread the word of Truth, to keep people from thirst and to offer salvation.

## 7. The Lives of Great People

I would like now to touch on the lives of great people. If we look back over the history of humankind we come across various great people. Of course, they were not all religious figures; outstanding people appear in all kinds of fields. They lived in love to guide others, embodying love itself, and their lives deserve serious contemplation. They were not content with love in its usual forms, the love that grows between men and women, matrimonial love, or the love between parent and child. In fact, they teach us that love can exist opposite to its normal forms.

Jesus Christ was one of these people. In terms of filial love, he was quite lacking and cannot be described as a good child. Born the son of a carpenter, he was expected to follow his father's trade and become a skilled craftsman, to marry and raise children to carry on the family line. But this was not to be. He was even rude to his mother, who was later to be revered as the Virgin Mary, saying: "Our souls come from heaven, not from other people; therefore, my lady, although you may be my mother physically, spiritually you are not. This is something that you need to understand." Looked at from his mother's point of view, this was a very cruel thing for him to have said.

He did not get on very well with his brothers either. He had four brothers but, unlike Jesus, they were quite ordinary; of all his family, only Jesus stood out. Neither his father nor his brothers could understand him. They could not see why he involved himself in some new religion and went around saying all kinds of strange things instead of helping in the family business. However, he lived for the love of a higher plane; he lived to save the whole of humankind. We must realize that there is a love that transcends domestic love, brotherly love, and filial love.

The same can be said of the life of Gautama Siddhartha, the man who came to be venerated as the Buddha. Born the son of a king, he fled the Kapilavastu palace at the age of twenty-nine, leaving behind a wife and child and turning a deaf ear to his father's entreaties. After that, he became an ascetic, spending six years in the mountains, disciplining himself as he strove toward enlightenment.

He was born a prince and expected to succeed his father as king, but he cast all this aside. So it is only natural that he be described as a bad son. He had a wife named Yashodhara and a son named Rahula whom he deserted, not returning to Kapilavastu until he had attained Buddhahood. In so doing, he destroyed matrimonial love while casting away his parental love. However, these were bonds that he was forced to break for the sake of a much higher ideal.

It would not have been possible for him to attain Buddhahood while living in the palace, and as a prince he would not have been able to teach the Law to the people. Things were very much different in those days, and we must not try to judge him by modern standards. It should not be overlooked, however, that after his Order was firmly established, he summoned his wife and child, making them disciples and looking after them while at the same time welcoming many other men and women of his clan (the Shakya clan) to join them.

I am not presenting the lives of Jesus or Shakyamuni Buddha as examples of how people in the present or future should live, denying love for their spouse or children. If it is possible, we should save the world while preserving the harmony of the home, the relationship of parents and children and our job, and this represents the universal form of love. However, I want you to realize that there are those among the great who expressed exceptional forms of love and their lives continue to light our path to the future like the sun through the power of "existence as love." We should all show respect to these people for the lives they led.

## 8. As the Embodiment of Buddha's Love

In the last section we looked at the examples of two great men who each ignored worldly love, recognizing their role to live out the love of a higher dimension. They sacrificed the things of this world for the sake of things of a higher dimension. In other words, to them the love for Buddha was far greater and far more important than the love for other human beings.

The difference between a life of love based on wavering, ever-changing human emotions and that of a love based on the eternally stable and everlasting will of Buddha is self-evident. Eternal love, truly unchanging love, is expressed through the effort to embody the will of Buddha, and through letting His will be our guiding principle. We cannot deny the importance of the love that exists between men and women, parental love, or brotherly love and should treasure it for what it is. But we should recognize that it is merely an instinctive form of love that we are provided with to prepare us for a higher form of love, the love for Buddha.

I would like now to consider both the "love that comes from Buddha" and "love toward Buddha." Buddha loves humankind with an infinite love that knows no bounds; but this love is not a love that desires love in return. It is not a love of give and take, it is an unconditionally giving love. In the same way that the sun provides Earth and all the plants and creatures on it with everything they need without asking for a penny in return, so Buddha gives His love unstintingly, shining as pure love, as the highest form of "existence as love."

We should all be aware of this love; we should all realize just how much love we receive. If we fail to give thanks for this constant supply of love, we can hardly describe ourselves as being children of Buddha. Surely it is shameful to receive so much wonderful love, day in, day out without a break and not experience any feeling of gratitude. However, there are many people in the world who pay no respect to Buddha for the infinite, free Love of Buddha. In fact, many of them do not even appear to be conscious of its existence.

We who are aware of this love, who realize that Buddha loves us, should try and do something to repay Him for it. When we become adult and have children, we offer them the same love our own parents gave to us. Buddha is both father and mother to humankind, and, as our parents, He offers us His boundless love. So surely we should find some way in which we can return this love to Him. In other words, we should live our lives conscious of the fact that we are children of Buddha. We should not offer our love to people in an effort to win praise or a good reputation; we should simply offer it in the same way that Buddha offers us His limitless love.

Looked at through spiritual eyes, people catch the love of Buddha in much the same way that a television antenna catches the signals from the broadcasting station; what I want to say is that as we receive so much love from Buddha, we should try to pass some of it on to others. We are all receiving the love of Buddha, so we should aim to circulate it. It is our duty to let the love flow through us and on to others in much the same way that a river flows from its source down to the sea.

## 9. Differences in Souls

I would like now to think about the differences that exist between souls. It could be said that the soul is a vessel designed to receive the love of Buddha. If the vessel is small, it will soon overflow; but a large vessel will be able to hold an abundance of the love of Buddha. In the same way that a reservoir holds a lot of water that can be released to turn a turbine and create electricity, our souls contain a reservoir of love that can be used to generate power, the amount of power depending on the quantity of love it can hold.

The largest reservoirs belong to the saviors of the ninth dimension. They are vast in size, and when they let their love flow out they create enough energy to supply the whole world. Therefore, the differences of souls can be measured by the amount of power they are able to generate when they release the love they store in their reservoirs.

When electrical power is generated by the power of falling water, the further the water falls, the greater its energy, and, therefore, the higher the reservoir is situated, the more power it is capable of producing. In a similar way, the higher the soul is situated, that is to say, the nobler its character, the greater the flow of love and the greater the amount of energy it can produce.

Accordingly, in order to catch as much love as possible, it is important that we focus our energies on "enlarging the vessel of the soul," and on "raising our character to the highest level possible." To enlarge the vessel of the soul, we must struggle to increase our broad-mindedness and tolerance, to cultivate a character that is so magnanimous that it can envelop and embrace everything. This is an important objective of spiritual training. Another way of achieving this is to raise the height of the reservoir, that is to say, to raise the level of our souls through assiduous and indefatigable work to come closer to Buddha. This kind of training is the way to enlightenment.

So what is enlightenment? What does the word mean? It is the "nourishment of the soul" that is acquired through the effort to absorb the Truth and put it into practice. This nourishment and the experience it brings is what is known as enlightenment. Therefore, I recommend that you should try to absorb and practice the Truth at all times, manifesting it in your daily actions and spreading love. In this way you will be able to increase the tolerance of your soul while raising its level to build a vast reservoir within.

## 10. That Which Surpasses Love

We have been thinking about the love that runs through the seventh dimension; but it must be realized that love is very much a relative concept. Love can exist between people, between people and animals, and between people and plants—but it cannot exist in a void, it cannot exist of itself. A jewel will shine of itself, it does not need anybody else; but this cannot be said of love. Love passes from one person to the other, it is a mutual giving; this is the true essence of love.

Having said this, however, I must add that a mutual, relative love is not the only form of true love. There is also a love that shines like a diamond or like crystal when the morning sun hits it, a love that sparkles even when alone. This kind of love far surpasses that which people feel toward other people, animals, plants, or even minerals. What is it that could be greater than love? There is only one answer; the thing that surpasses love is compassion. A diamond sparkles in the light, but not because it expects something in return, it simply shines eternally. In the same way, there is a love that shines of itself, without expecting anything in return, a great love that surpasses all other forms of love.

Sometimes when we are walking in the mountains we may come across an azalea or violet blooming among the rocks. Why does the flower look so beautiful? Why do flowers bloom? Flowers bloom because that is the reason for their existence and they present us with questions on the importance and value of existence. Lilies bloom in the valleys for the sake of blooming, diamonds shine for the sake of shining—this is what is meant by compassion. Compassion has value of itself irrespective of the existence of others. Compassion is something that transcends love.

Compassion is "existence as love." Existence itself becomes love, a person only has to be present to provide love. This form of love represents a state close to Buddha. Buddha, through His very existence, provides love for all creatures. Compassion or "existence as love" is what surpasses the love that is experienced between human beings and, for this reason, it is what we must all aim to attain.

# Five: The World of the Eighth Dimension

## 1. What Are the Nyorai?

In the proceeding chapters I have given an overall picture of the spirit world up to the Bosatsu Realm and in this chapter I would like to move on to cover the Nyorai Realm of the eighth dimension. Like the word Bosatsu, Nyorai is a Buddhist term. It is roughly the equivalent of archangel in Christian terminology, and is also sometimes referred to as Great Guiding Spirit of Light. The word "Nyorai" is written in Japanese with two characters, "nyo" meaning "as if" and "rai" meaning "come;" but what do the two characters put together mean? It means "as if coming from the Truth"—that is to say, it is the embodiment of the absolute truth that has come down to Earth from the highest realms of enlightenment. It is difficult to generalize about the status of individual Nyorai, but suffice it to say that they have been incarnated on Earth and have been recorded in history as some of the greatest people of all time.

Some of you may be wondering how many Nyorai there are altogether. It has been estimated that the total population of the Real World is about fifty billion, but of these less than five hundred are Nyorai. There are as few as that. This works out to an average of approximately one Nyorai per one hundred million people. The population of Japan is presently one hundred and twenty million and so statistically one could expect there to be one Nyorai among them.

However, at times in history when a major advance is due to be made in the teaching of the Law, the Nyorai tend to appear in greater numbers, and so it is impossible to say just how many there will be on Earth at any one time. Whatever the period, though, there are never more than a few Nyorai living on Earth at the same time. It is impossible for them to appear in their tens or hundreds. In a way they can be said to resemble pinnacles. Standing alone, Mount Fuji looks so impressive, but if the whole of Japan were covered with high mountains like Fuji, the resulting scenery would not be so beautiful. The Nyorai are those who stand out among their contemporaries, like Mount Fuji.

One period when Nyorai appear in large numbers is when a major civilization is reaching its peak. For instance, if we look at the ancient Greek civilization, Socrates was a Nyorai, as was his disciple, Plato, and Plato's student, Aristotle. Another Nyorai who lived in Greece at about the same time was Pythagoras who was followed later by Archimedes (of the ninth dimension). The early Chinese civilization saw the appearance of Confucius (of the ninth dimension), Lao-tzu, Chuang-tzu, and Mo-tzu, who were all Nyorai. Jesus Christ (of the ninth dimension), and John the Baptist were two Nyorai, while among the prophets of the Old Testament, Jeremiah and Elijah were also Nyorai. Of course, Shakyamuni Buddha (of the ninth dimension) and several of his followers who founded the Buddhist religion were also Nyorai.

It can be seen that the Nyorai appear on Earth to raise a particular culture or era to its peak. They form a nucleus for the study of the Law, or become active leaders in cultural or artistic fields, allowing the era to reach new heights. Eventually the culture they created falls into decay, and then various Bosatsu come down to breathe new life into it. When finally it deteriorates beyond repair, the Nyorai again appear to create a new culture, and the cycle is repeated.

## 2. The Nature of Light

When you enter the eighth dimension, you will inevitably become conscious of light. Light is a word that we use in a variety of ways, saying: "Buddha is light and the essence of human beings is also light" or that "The high spirits carry out their missions through the seven-colored Light of Buddha." But what do we mean by the term light? Do we mean something like the light of the sun? We tend to use the word "light" indiscriminately and so now I would like to present a proper definition of the meaning of the "Light of Buddha." What do we mean when we say that "Buddha is light?" One way of defining the concept of light is to contrast it with its opposite state and thereby accentuate its qualities. Of course, the opposite of light is darkness, and when we think about the attributes of darkness we can say that it is gloomy and treacherous, offering no hope and no vitality. If this is so, then light must mean the opposite.

If we were to list its qualities, we could say that light is bright. It is the source of the energy of life, which contains various intentions and wills, characters and natures. However, when we talk about light and dark in this way, we come up against the old debate of monism versus dualism, that is to say, whether darkness exists of itself. The fact is that darkness is a passive existence. Nighttime is not created by rays of darkness hitting the Earth;

darkness is only created when something cuts off the rays of light. On the other hand, light is a positive existence, an active power.

No matter how strong a light may be, if something stands in its way darkness will be created. The stronger the light, the deeper the darkness. For instance, even if a light is ten thousand candlepower, or a million candlepower, it will not be able to illuminate something that is in the shadow of a rock—that is the way light behaves. Its nature is to move straight ahead, so if something obstructs its way, it is cut off.

The same can be said of good and evil. Good embodies a positive existence, whereas evil is negative. However, it is not necessarily true that only good exists and evil cannot. Although evil is basically a passive existence, it can come into existence through the intervention of a third party. Darkness always appears where there is light. In the same way, although evil did not originally exist, it appears wherever there is good in order to make the good stand out. While it is true that evil is not a positive existence and simply represents a lack of good, a lack of goodness can act as a medium through which evil can exist.

Imagine that you illuminate your home with an extremely bright lamp. Regardless of the power of the light, you will not be able to avoid shadows forming somewhere in the room. Of course, if the room were walled with mirrors there would be no shadows; but in a normal house shadows are sure to form no matter how well lit it may be. Crockery or furniture will block the light and create shade. Therefore, we can see that although shadows, darkness, and evil do not exist of themselves, they come into being in the natural course of human living.

### 3. The Essence of Space

Having explained the Light of Buddha, I would now like to approach the problem of space, while continuing our exploration of the nature of light. What is space? This is a question that has long puzzled humankind. Space may be described as having height, width, and depth, making it three-dimensional like a box. This is one definition; but, unfortunately, it is not entirely correct. To say that space is three-dimensional is to ignore the fourth, the fifth, the sixth, the seventh dimensions, and the subject of this chapter, the eighth dimension. This is something that needs bearing in mind.

The essence of space is a "field" where a consciousness intends to make something exist. It is a field where phenomena take place. In other words, it is a field that energy can pass through, a place where energy can perform as it should.

This is the original meaning of space. It is not simply a three-dimensional cube but a place where the Light of Buddha can be active and perform various operations. Therefore, in the fourth dimension and above, our three-dimensional definition of space does not apply. Multi-dimensional space is a field of consciousness where the Light of Buddha can create various phenomena and carry out its diverse activities.

## 4. Eternal Time

Time is a concept that is often compared to space, and time is the subject I would like to contemplate in this section. It is often said that space represents a horizontal expansion whereas time is a vertical expansion but this gives rise to the questions "can space exist without time?" and "what is the relationship between time and space?"

In the last section I gave the definition of space as being the field where the Light of Buddha can be active. Light being active infers movement of some kind, and movement in turn requires a certain passage of time. Therefore, if time were to stop, what would happen to space? Would light manage to remain active if time were to stop or would it too be stopped? The answer is that when time is stopped, light is also frozen, like a single point in a photograph. In order for space to carry out its primary role acting as a field for light, it is necessary that time be inherent within it. We can say that "space cannot exist without time," "time and space cannot be separated; it is the presence of time that allows space to function as such," or "it is the continued existence of the field that allows light to act."

You should not look upon light merely as rays of visible light; it can be broken down into minute particles called photons. These photons combine to produce tiny particles that are the building blocks of creation. Everything in the universe, ourselves included, is made up of these particles and so all material objects can be said to be made of light. Some light coalesces to create material objects while the remainder makes up the spirits and spiritual energy of the fourth dimension and above. This means that not only the third dimension, but also the fourth dimension and above—everything in the universe—is created of light. Therefore, since everything is made of light, and space is a field where light can be active, it follows that if there is no place where light can be active, space does not exist. This means that the space of the third dimension and above consists entirely of light and the activity of light.

Without time, light is unable to be active, which means that without time there would be no space, no material objects, no spirits, nothing. There would just be a floating mirage-like space, and there would not be real space, the space where light could be active. Since space only exists as space in order to provide a field for light to be active, it can be argued that time is an important component in the existence of space. Looked at in this way, it can be said that the world that Buddha created, that is to say, the third, fourth, fifth, sixth, seventh, and eighth dimensions, consists of space that encompasses time, and light that moves within it.

When Buddha created the world, He did so out of three elements: light, space, and time. Light is transformed to create the material and spiritual bodies of all the dimensions, while it needs space as a field in which to be active. The passage of time is also necessary in order for light to be active, for light to flow as light, for light to reach its goal.

## 5. Signposts for Humankind

When we understand the essence of the world that surrounds us, we become aware of why we are alive, how we should live, and what the objectives of life are. Seeing the overall picture of the world that Buddha created and understanding its essence means to discover signposts that will guide us through our lives.

What are these signposts that will guide us through the journey of our life? You can find them through understanding that all human beings are allowed to live in the world that Buddha created with light, space, and time, and through comprehending the objective of the creation of the world.

So, what kind of world did Buddha intend to create, combining the three elements of light, space, and time? First, let us imagine that space is like a transparent glass box and that a beam of light shines in from one corner. The light is reflected off the inner surfaces of the box, bouncing forever around the interior. The light is trapped within the box and moves around to create all kinds of scenes. It is an art form that consists solely of light. The light hits one wall but is reflected to the next then on to the next, until the interior of the box is filled with a movement of light. When we look at the history of the universe and humanity, with this kind of perspective in mind, we realize that light is not merely the result of chance, but is following a purposive course. In other words, the light that is emitted by the Primordial Buddha is not directed at random, but with the clear objective of furthering the development of both the universe and humankind.

The aims of the activity of Buddha's light can be summarized into two points, the first being "evolution." If we look at the universe and Earth, the history of Earth and the history of humankind, the great purpose and objective of evolution is to be seen within them all. This is a fact that nobody can deny. Only because we strive to attain a higher level of development are we given life. This is the only thing that makes our lives worthwhile. If humankind simply lived in degeneration, then we would have to question the very meaning of our existence. Why should we exist at all if only to degenerate? Working with clay is a pleasurable pastime, everyone derives satisfaction from making something out of nothing, but if animals and people were made of clay, and started life only to end as blocks of clay, it would be completely meaningless. To make something out of nothing, this is the essence of evolution. To make something out of nothing, then to develop it into something even greater, is one of the objectives of humankind.

The second aim is to create "harmony," a great and grand harmony. So what do I mean by great and grand harmony? Imagine that Buddha created a mountain of clay within a vast room, then from this clay He made the Sun, the Earth, the Moon, then plants, animals, people and all kinds of other things to fill it. It is a wonderful thing to make things evolve from shapelessness into form in this way, but the next question is whether these things He has created co-exist in an orderly and beautiful way. Is there balance between plants, animals, and human beings, between the Sun, the Earth, the Moon, and all the other planets and stars He created? The equilibrium between day and night, between land and sea, between hot and cold—this harmony was the next aim that Buddha set for Himself. Looking at the history of humankind, it can be said that there has been passage of time centering around the two main objectives—evolution and harmony, or development and harmony.

## 6. What Is the Law?

I have explained that the signposts for humankind are evolution and harmony, and now I would like to think about the "Law" that we have been researching and seeking throughout the ages. What is the Truth and what is the Law? How are they explained as systematizations of Truth? What role do they play?

The Law also comprises the two elements of evolution and harmony, which I explained in the last section. These two elements are contained within the Law. The Law represents the rules of the universe; it is the systematization of the Truth, and its objectives are evolution and harmony. Personal improvement, the development of the individual, is an integral element of the Law. Any law that does not work for the improvement of the individual is not a true law; it is not part of the Truth of Buddha, it never has been and never will be. The principles of progress and evolution are enshrined within the Law and so at the very least it will contain provision for the maturity, heightened enlightenment, and improvement of the individual.

The resulting improvement of people in general is good, but sometimes this can lead to conflict of the freedom of individuals. For this reason it is necessary to have a law that resolves such conflicts and leads to the improvement of a community, a congregation of individuals. For instance, let us say that there is somebody working for a company who feels that he is qualified to be president, and there are two others who also feel that way. However, all three cannot hold the post at once. Then, the members of the board weigh up the individual merits of all three to see if any of them have the skills necessary to run a company of several hundred or even several thousand people. If one does, he will be given the post and the other two passed over. However, if all three applicants appear suitable for the job, then an order is established with A becoming president first, to be followed by B then C. In this way, the development of the individual is balanced against the good of the group.

Numerous religious leaders, moralists and philosophers have appeared throughout history to teach this principle of adjustment and harmony to the world. China's most famous philosopher, Confucius, espoused the seniority system and said that the young should give precedence to their elders. According to his philosophy, if the three people who wanted to become president were equal in their abilities, then the job should be awarded to the eldest. This system still exists, to some extent, to this day. Even though somebody's age does not necessarily correspond to their spiritual development, people still tend to feel that a person's greater experience will lead to greater wisdom, and this gives grounds for the seniority system.

This system does not suit everybody, however, and another school of thought is represented by the merit system. In this, the applicants are given a test or their past achievements are compared, and the one with the highest results is offered the job. Yet another school of thought is utilitarianism, as put forward by Jeremy Bentham, which extols "the greatest happiness for the greatest number." John Stewart Mill was also an exponent of this movement that bases a choice on what will benefit the most people. I think that this way of thinking makes sense.

Basically, some kind of principle is necessary in order that the development of the individual be of benefit to society as a whole. Actually, this way of thinking encompasses the philosophy of both the Lesser Vehicle (Hinayana) and the Great Vehicle (Mahayana) of Buddhism. The Lesser Vehicle, which focuses on the enlightenment of the individual, embodies the principle of progress, whereas the Great Vehicle Buddhism, as a methodology for building a utopia in this world, incorporates the principle of harmonizing conflicting needs.

Thus the Law contains two main principles, progress and harmony, and it is through the fine balance of these that humanity as a whole can achieve happiness.

## 7. What Is Compassion?

I would like now to think about the Law and compassion. At the Institute for Research in Human Happiness, based on the principles of individual progress and social harmony, I teach that the pursuit of happiness is part of human nature and that this happiness for which we seek is twofold—private happiness and public happiness. By private happiness I mean the pursuit of each individual's happiness, while by public happiness I refer to the spreading of the personal utopia that is achieved through the pursuit of personal happiness to envelop society, the world, and humankind as a whole, thereby creating a public utopia. This is the fundamental teaching of our movement. The question now is, why should it be necessary for us to pursue private and public happiness in this way? Is there some guiding principle behind this search?

As I said at the beginning of this section, it is in human nature to search for happiness, the reason being that Buddha bestowed this penchant on us through His compassion. He gave a purpose to our lives, and if that purpose was to lead to unhappiness, the world would be a wretched place indeed. However, Buddha implanted the desire to strive for happiness in the soul of each of us in order that we should all be as happy as possible. This inclination is embedded so deeply within us that it has become part of what makes us human.

You may wonder why Buddha made us in this way, and the answer is that we are all children of Light who have split off from the Primordial Buddha Himself, and, as such, we carry within us the same nature. So what is the nature of Buddha? It can be said to derive great pleasure from happiness that is born of development and harmony. The reason why Buddha is a great force that controls and governs the universe is that He contains within His being the energy of happiness. Therefore, we can argue that the reason why Buddha exists as Buddha, the very purpose of His existence, is based on happiness.

So what is it that gives Buddha happiness? What makes Him feel joy? The answer to this is in the process of producing, nurturing, developing, and bestowing prosperity on all of creation in a flourishing harmony. He feels joy while experiencing this process. If Buddha were to cease to exist as Buddha, then there would be no more joy. However, Buddha is active as Buddha, and His aim is to bring harmony to all creation, allowing it to develop and prosper, and in the process, He has magnificent, beautiful experiences that give Him joy and pleasure.

In this way, Buddha Himself undergoes a great transformation, and attains even further development of Himself. The facts that "humanity exists to pursue happiness" and "people were created to live in joy," represent the true nature of Buddha and are in themselves proof of His great compassion.

## 8. The Functions of the Nyorai

In this section we will look at the function, role, and work of the Nyorai. In *The Laws of the Sun* I stated that Buddha made all souls equal, but that is not all; He also evaluates them fairly according to their achievements. By fairly, I mean that those whose task is to lead others are awarded a suitable position, role, and strength to achieve their goals. The superiority of the Nyorai is based on this principle of fairness.

Although we are born equal as children of Buddha, those who have succeeded in accumulating wisdom through repeated incarnations on Earth are rewarded with a suitable position to assist them in achieving even greater self-realization. Nyorai function as representatives of Buddha. Buddha does not take a human form, Buddha is the being who creates this vast, multi-dimensional universe in which we live and it is beyond our ability to see Him. The Nyorai, therefore, act in His stead, presenting a mighty countenance in which we can feel the greatness of Buddha. In other words, the Nyorai exist in order that humans may be able to feel the existence of Buddha. This is the basic reason that Nyorai are said to embody "existence as love."

The Nyorai are existences that "come from Truth." They are the embodiment of Absolute Truth, the beings who manifest love toward humans. In other words, the Nyorai are beings whose very existence leads us to the realm of enlightenment, illumination, and happiness. Therefore, the Nyorai can be described as Light itself, the personification of Buddha's light.

It is impossible for humans to see, understand, or comprehend Buddha Himself; however we can obtain an analogical inference of Him through the Nyorai. They exist to present us with a metaphor for Buddha and allow us to conjecture upon His compassion. Although we may not be able to see Buddha directly, we can perceive His great compassion, and understand His power through the existence of the Nyorai. We can sum up by saying that the Nyorai exist to instruct us, that in themselves they represent an education for all of humankind.

### 9. Talking of Buddha

Ultimately, the Nyorai's role is to explain what Buddha is. The Nyorai exist, as Buddha's representatives, to instruct us in the meaning of Buddha. Ordinary people are not in a position to talk with authority about Buddha, but the Nyorai are much closer to Him and therefore better able to carry out the role of teaching what Buddha is. As Great Guiding Spirits of Light, the inhabitants of the eighth dimension have been permitted to speak of the many attributes of Buddha.

However, Buddha is so great that even the Nyorai are unable to teach everything about Him. The soul of a single human being, no matter how advanced, is incapable of grasping a subject so vast. Therefore, the Nyorai of the eighth dimension are divided into separate groups, each teaching the aspects of a single color of the Light of Buddha.

The Nyorai who are connected to the yellow (or golden) light that is controlled by Gautama Siddhartha concentrate on the aspects of enlightenment, the Law, and compassion when describing what Buddha is. The Nyorai who concentrate on the white light that is controlled by Jesus Christ talk about Buddha from the standpoint of love. Those of the red light that is governed by Moses try to teach what Buddha is through the miracles that He performs. The Nyorai of the other colors also work in the same way. For instance, the Nyorai of the green light, as represented by the Chinese philosophers Lao-tzu and Chuang-tzu, teach about the great harmony of nature. That is to say the Nyorai who receive the green light preach that Buddha can be seen in nature and in the harmony that exists there. Zeus is in charge of the light of art, and the Nyorai in his charge tell us that Buddha can be seen in the work of artists. Confucius controls the violet light, which teaches propriety, order, and loyalty, saying that the way to Buddha requires reverence and worship, thus trying to indicate what Buddha is.

Looking at the work of the groups of Nyorai, it can be seen that while each Nyorai is basically teaching the same thing—talking about Buddha—they do so through the influences of the particular color or light to which they are affiliated. However, humankind has never realized that these colors with various attributes are all simply different aspects of the same Truth, and, as a result, the history of this world has been full of wars stemming from differences in religion. People could not see the different roles of the Nyorai, denouncing the followers of different philosophies as heretics and striving to destroy them. However, the time has now come when people should look at the teachings of the Nyorai representing all the different colors of light and understand what Buddha really is.

## 10. The Path to Perfection

I have spoken of the Nyorai as something much greater than the rest of humankind. But does this mean that they have completed their spiritual training? Is there nothing left for them to learn? This is the point I would like to think about in this section.

The truth is that with regard to being reincarnated on Earth and undergoing training here, they are still very human. While they are in the eighth dimension they are great beings, each a specialist and the embodiment of one of the colors of spiritual light. But when they come down to Earth every few hundred or few thousand years in the cycle of reincarnation, they experience the life of human beings, seeing and hearing a multitude of things. While they are living on Earth they are exposed to other points of view and able to learn things that belong to different colors of light. In this respect, it can be said that, even though they may be Nyorai, they are still working toward total enlightenment, although it cannot be denied that they are infinitely closer to perfection than the rest of humankind.

So what is the path to perfection that they are striving to attain through their training? The answer is that they devote themselves to their training in order to gain a higher perspective of humankind, the Law, the world, and history, with the aim of realizing a greater integration and synthesis. In other words, they want to deepen their insight and their understanding of the world.

Buddha ordained that humans should experience reincarnation as a way of evolving and developing themselves, and no soul is exempt from this rule. Many people believe, however, that the Nyorai have managed to escape from this fate, that they are no longer bound by these laws, and I think this point needs some clarification. Ultimately, not even the Nyorai are exempt from this rule, they cannot remain in the other world indefinitely, but the big difference is that they are able to choose when and where they wish to be reincarnated in accordance with their own plans. The Bosatsu and souls of the lower dimensions are sent down to live on Earth as necessary, being instructed to live in various ages as part of a compulsory education. However, the Nyorai have finished this part of their schooling and are like adults, free to choose what and when they would like to study. In this, they are like university graduates who have finished a set course of education and are free to continue their studies in their own way. The goal they have set themselves is to achieve a deeper insight and understanding of the world and its people, to learn to see things from a more global, more universal standpoint. The Nyorai, too, are treading the path toward perfection through their own spiritual discipline.

# Six: The World of the Ninth Dimension

## 1. Beyond the Veil

In the previous chapters I outlined the worlds from the fourth to the eighth dimensions, explaining the laws that govern all these realms. Looking back over history, I think I can truly say that there have been very few other people to have done so in such detail. Moreover, what makes this book unique is that in this chapter I will offer a description of the ninth dimension, which until now has remained hidden from the eyes of philosophers and men of religion. As much as possible, I intend to explain this mysterious and obscure realm in a way that is comprehensible to we who live on Earth.

The world of the ninth dimension, which has been veiled from sight until now, is the realm of saviors. The inhabitants are beings who are considered saviors or messiahs and who only come down to this world once every thousand years or more. Each civilization has its unique characteristics, and humanity has experienced infinite number of eras with a wide variety of attributes.

Depending on an era, some spirits from the ninth dimension may decide to descend to Earth while others do not. A savior may become incarnated every two or three thousand years while a civilization lasts, but he may not appear at all in another civilization. The inhabitants of the ninth dimension have separate roles and they decide who will share responsibilities during a particular era to give each civilization a unique character.

Among the saviors who are most famous in contemporary civilization are Gautama Siddhartha, Jesus Christ, and Moses. Another well-known being of the ninth dimension, although he is not usually referred to as a "savior," is Confucius. One thing that all these people had in common while they were incarnated on this world is that they laid down the principles that were to become the basis of a particular civilization.

## 2. The World of Mystery

The ninth dimension is a very mysterious place. When we think about the Real World, we of the third dimension summon up a variety of images; but, when it comes to the ninth dimension, the only thing that we can say for sure is that its inhabitants are no longer completely human. The inhabitants in the Posthumous Realm of the fourth dimension have assumed their astral bodies but their lifestyle varies very little in appearance from that which they experienced on Earth. The same is true of the people who live in the Realm of the Good in the fifth dimension. They still think of themselves as human beings and many of them busy themselves with terrestrial professions. Farming is one popular profession in the fifth dimension and you may also see carpenters, school teachers, retailers, and mechanics. In this way, the fifth dimension is still easily comprehensible to our way of thinking.

By the time they reach the Realm of Light in the sixth dimension the consciousness of the spirits has been raised to such a level that they emit light in a godlike way but they still retain bodies that are quite human in appearance. Sometimes, however, they remember that they are in fact spiritual consciousnesses and behave as such, flying wherever they want at will. The spirits who had lived in the West while they were on Earth take on wings like angels whereas those from the Far East may travel on small clouds, as depicted in old stories of that region. In this respect, they can be said to differ slightly from terrestrial beings whose perception is governed by the five sensory organs.

In the Bosatsu Realm in the seventh dimension the inhabitants still adopt human form in order to continue their training. However, since a major part of their work concerns the education of people in their developing phases, they do not spend all their time comfortably in the seventh dimension, but visit the sixth, fifth, and fourth dimensions to undertake various tasks or act as guiding spirits for those on Earth, in order to make the world a better place. Playing such versatile roles, these spirits of the seventh dimension no longer follow a human lifestyle. In other words, they have a higher degree of self-awareness and self-understanding. However, when they want to look at themselves objectively, they tend to do so in the form that they used when they lived on Earth as humans.

When we reach the Nyorai Realm of the eighth dimension, things are a little different. The Nyorai sometimes appear on Earth in their role as guiding spirits to teach religious leaders, and when they do they will take on a godly appearance. For instance, when the Shinto god Ame-no-Minakanushi-no-Mikoto (the Lord God of the Heavenly Center) manifested himself as the Great God of "Seicho-no-Ie" (a modern Japanese sect), he did so in the form of an old, white-haired man. However, when they are in the other world, the Nyorai do not assume human form in their daily lives. They no longer feel the need for it. Sometimes they may take on human form to facilitate conversation or the exchange of ideas with others of their own kind, but this is not their usual state. They are also able to divide themselves into more than one corporeal body and take on any appearance they may desire. Nyorai may take a part of their consciousness, split it off into an independent form, and use it to achieve a variety of goals. In this way they can divide their beings into numerous rays of light, creating as many entities as necessary to achieve a particular aim.

So far we have covered the realms up to the eighth dimension. When we arrive at the ninth, however, things become much more mysterious and harder to understand from a terrestrial point of view. I have said that ten people live in the ninth dimension. But, rather than think of them as people, it would probably be more accurate if we were to think of them as ten gigantic bundles of rays, each with its own individual characteristics. When they want to communicate with me here on Earth, they take on the appearance and personality they used in their previous incarnation, but they are not usually to be seen in this form.

This is a very difficult subject to try and explain, but I will attempt to do so using electricity as a metaphor. Let us say that there are ten batteries in the ninth dimension, each with its own particular characteristics. Each battery has a wire attached to its positive and negative terminals with a variety of different light bulbs in between, so when the electricity flows the bulbs are illuminated. Each of these bulbs has a different name. For instance, one may be called La Mu, the next Rient Arl Croud, then Hermes, then Gautama Siddhartha. Although there may be several light bulbs, they are all connected to the same one battery and all use this same electrical source. When necessary, such light bulbs are lit to express different characters.

### 3. The True Faces of the Spirits of the Ninth Dimension

So what is the true form of these inhabitants of the ninth dimension—these beings we call gods? Do they really sit on a throne in a palace, wearing a crown and white robes as they are depicted in books? No, they do not. The spirits of the ninth dimension are better described as electromagnetic forces, energy bodies, or consciousnesses, and it is in this form that they carry out their activities. When such a being wants to be recognized by a spirit from one of the lower dimensions, it illuminates a light bulb and by so doing makes itself visible. I have said that Jesus is an inhabitant of this realm, but that is not to say that he lives there in the form with which we are all familiar; the form of a thin man on the cross with long hair and a beard. There is only a mass of light incorporating the characteristics of Jesus that works to instruct the people on Earth or the spirits of the eighth dimension and below when necessary.

When Jesus appears to the Nyorai of the eighth dimension or the Bosatsu of the seventh dimension he does so in the form he took when he lived on Earth, as this makes him easier to recognize. However, the only spirits capable of seeing him in this guise are those who dwell in the eighth, seventh, and perhaps sixth dimensions. To the inhabitants of the lower dimensions he will simply appear to shine like a dazzling light and they will be unable to make out the details of his form. In other words, the volume of light is that great in comparison.

I have spoken at length about the various dimensions. But what it all boils down to is that the inhabitants of the different realms differ from each other in the amount of light they emit. When I speak of light, I do not mean light in the way we on Earth think of it. It has inherent qualities and it looks as if there are bundles of yellow, white, red, and green rays. Of course, when I speak of yellow, white, red, or green, I am only using words that can be understood here on Earth, their true form is something quite different.

On Earth, color, in its true sense, does not exist. What we think of as blue is not really blue at all, it is merely a surface that reflects the blue rays within the sun's spectrum. Something that absorbs all the sun's rays appears black to us, while something that reflects them all appears white. Something that reflects only the yellow rays appears as what we call yellow. In this way, nothing really has any color of its own, it is merely the particles that make up the object reflecting a certain color that makes it appear colored. It is easy to prove that color is essentially non-existent. If we turn out the light, nothing retains its color. If a color really existed, we would be able to see it in the dark, but be it red, white, or yellow, colors all disappear in the dark. The truth is that what we think of as colors are simply reflections of certain wavelengths of light and they disappear when the light is cut off. All that exists are various wavelengths of light that when reflected appear to us as colors.

## 4. The Essence of Religion

I would like now to consider the essence of religion. In my description of the eighth dimension, I said that the Light of Buddha is split as if put through a prism, the teachings of each of the separate colors being preached by individual Nyorai. They each taught their own ideas of Buddha (or God), and these teachings developed into major religions of the world. But why should it be necessary to divide religions in this way? Surely it would be better if Buddha were to present a single religion and allow all the religious men in the world to work together. In this way it would be possible to avoid all the confusion, the wars, and the chaos that exist today.

While this idea seems very good on the surface, it is mistaken and contains many hidden dangers. If humankind were presented with a "ready-to-wear," "one size fits all" religion, would we really be satisfied? We cannot even make up our minds about the kind of car we want to drive. Each of the car manufacturers produces different styles, each available in white, red, yellow, blue—every color under the sun. Cars can also be categorized by size, degree of fuel-efficiency, and price. On top of all this, there are new cars and secondhand cars and everybody chooses the one that suits their budget and needs.

I have only chosen cars as an example, but why should there be so much variety in the car market? It is because a car represents more than just a means of transporting people or freight. If cars were merely tools to take people from one place to another, there would be no need for all this diversity. But they also play another important role—they act as symbols of their owners. Cars are used as a way of advertising their owners' financial or social status. They are also a way of representing the owner's personality. They show if he or she is practical, or more interested in a car as a status symbol. There are numerous reasons for choosing a particular car. There are cars that appeal to men and those that appeal to women. Speed may not necessarily be a criterion; some people like a fast sports car while others might be quite happy to drive a slow car. Finally, there is the question of design. Some people like two-door cars while others prefer four-door or even five-door cars.

In this way, it can be seen that, even with something as simple as a car, it is impossible to say that any one particular model is the best; so how much more must this apply to religion? There are a plethora of religions in the world today, but we can no more say which religion is true, any more than we can say which car is best. Of course, we can generalize and say that the more expensive the car, the better it is, or that one car is of higher quality than another. But this does not necessarily mean that it will appeal to everybody. Taste varies from person to person.

Buddhism includes the ideas of both the Lesser Vehicle and Great Vehicle schools. Although people use these terms without really thinking about the meaning, the word "vehicle" in both of them really does refer to a vehicle as we understand the term. Lesser Vehicle Buddhism can be thought of as a small car that can only carry the driver whereas Great Vehicle Buddhism is like a bus that can carry a large number of passengers.

Thus, it can be seen that religions, like vehicles, range from the compact to the large, and the difference between them is basically how many people they can carry or in what way they carry them. For instance, nobody would want to drive a bus instead of a car for their everyday errands. Of course, a bus can carry a large number of people, but it is not suitable for a single individual's use. In the same way that there are various vehicles to suit various people's needs, religions also adapt themselves to people's tastes, climates, or environment.

A religion that sprang up in the Bible lands, where war and destruction were commonplace, required a god of judgment who would teach justice to the people. In the temperate climate of the Orient a god that would teach harmony was necessary. In the rational Western civilizations, the teachings took the form of philosophy. Whatever form it may take, the object of religion remains the same—that is to transport people from one place to another. The vehicles that carry them may vary, but in all of them we can discover joy and sense of purpose. This is how it has been ordained.

## 5. The Seven Colors of the Spectrum

I teach that the Light of Buddha consists of seven colors and this is indeed the case. The light is divided into these seven colors in the ninth dimension then transmitted to the Nyorai in the eighth dimension where it is split still further to produce ten or even twenty different colors. The ninth-dimensional beings who are responsible for the seven colors of light are as follows:

The central color, yellow, has a golden hue and is under the command of Gautama Siddhartha, otherwise known as Shakyamuni Buddha. Yellow is the color of Law or of compassion.

White light is under the command of Jesus Christ and is the color of love. The spirits of healing also fall under the control of this white light. Although it may only be coincidence that doctors and nurses wear white, it would seem to hint at the fact that they are under the influence of the spirits of the white light.

Moses is responsible for the red light. This is the color that governs, so world leaders fall under the influence of this red light. Red light is also referred to as the light of miracles, and whenever miracles or inexplicable phenomena occur they do so through the power of this light.

Next we come to blue light. This is the color of philosophy and ideology and, unlike the others, it is not controlled by one spirit but two. First, we have Zeus, who once lived in ancient Greece and is largely concerned with the control of art and literature. Art also falls under the influence of green light, but part of it is governed by the blue. The other spirit in charge of the blue light is Manu, the progenitor of the human race according to Indian mythology. He gives his name to the "Laws of Manu" (or Manu Smriti), which stipulate the daily conduct of the Brahman caste. He is a spirit of the ninth dimension who deals mainly with ideological matters, but also concerns himself with many other activities. At present, he is busy working on a unification of ideologies and beliefs to transcend race and nation.

Silver light is the light of science and the modernization of contemporary culture. This color is under the control of the spirit who was previously incarnated as Isaac Newton and before that as Archimedes. In this way, whenever he appears on Earth, he does so in the guise of a scientist, and while he is in the ninth dimension he works to promote scientific advance in the third dimension and above through control of the silver light. Thomas Edison and Albert Einstein are both Nyorai from the eighth dimension who worked through the silver light.

Next we come to the green light, which is the color that governs harmony and is the color of Lao-tzu and Chuang-tzu's ideology, the color of nature. This color is controlled jointly by Manu, whom I have already mentioned, and Zoroaster (Zarathustra), a prophet of a Persian religion, which believed in the ethical duality of good and evil and used fire in its worship. They work to teach the way of nature, the structure of the universe, and the harmony of all things.

Finally, we come to the violet light, under the control of Confucius, which deals with morality, a scholarly way of thinking, propriety, and order. In other words, it places things in hierarchical order and promotes a system of control based on seniority. The gods of the Japanese Shinto religion are, in fact, under the influence of this light.

I have now introduced you to eight Grand Nyorai and explained how each of them takes charge of a particular color of the Light of Buddha. However, I have already stated that there are ten Grand Nyorai who live in the ninth dimension. So who are the other two and what do they do there? It is important that I tell you this as well. One of them is called Enlil, but was known in biblical times by the name of Yahweh and served as the god of the Israelites, while in the Orient he was feared as the leader of the gods of disaster. The other Grand Nyorai is Maitrayer whose role is that of regulation. He is responsible for the splitting, adjusting, and balancing of the Light of Buddha, in terms of the comparative strengths of each color.

## 6. The Work of Shakyamuni

The central being of the ninth dimension is the one who was once incarnated in India as Gautama Siddhartha, otherwise known as Shakyamuni Buddha. However, when he did appear on Earth, he only possessed one fifth of the actual power he enjoys in the ninth dimension in the form of Buddha Consciousness (El Cantare Consciousness). From this you can imagine how vast a life force this consciousness is in the Real World. The origins of this spirit are extremely old and he is the most venerable being on Earth. One of the reasons why he has had such a powerful influence over humankind is his long history of working for the flourishing of Earth since its creation. Despite being the oldest spirit, he is still very active, sending part of his consciousness down to Earth on numerous occasions in order to instruct humankind.

As the eldest of the spirits, he is also responsible for the terrestrial spirit group and as a result it is no exaggeration to say that his character is reflected in all the various cultures that have bloomed on Earth. As I stated in *The Laws of the Sun*, he has lived on Earth numerous times, as La Mu of the Mu Empire, Thoth of the Atlantis Empire, Rient Arl Croud of the ancient Incan Empire, and Hermes in ancient Greece. The central task with which he is involved is that of creating the Law. If we look at the various religions, philosophies, and ideologies that have sprung up on Earth, we find that they can all be traced back to Shakyamuni. In other words, the things that he thinks about in the other world manifest themselves on Earth in a variety of ways.

The central core of what we refer to as the Shakyamuni Consciousness is called, in the ninth dimension, the El Cantare Consciousness. If we trace the Law back to its origins, we find that it eventually leads to El Cantare Consciousness. Thus the essence of the Shakyamuni consciousness is the Law that governs all of humankind.

## 7. The Work of Jesus Christ

I would like now to talk about Jesus Christ. He has been active since the creation of the terrestrial spirit group and is so famous that there is little need for me to tell you about his work. Jesus' mission is based on love. Now love has become a universal theme, spreading well beyond the boundaries of the Christian world, and this very fact proves how great is the power he wields.

Jesus is a consciousness of the ninth dimension who is also known by the name of Agasha Consciousness. Agasha was a Great Guiding Spirit of Light who lived in Atlantis toward the end of that kingdom's power and so influential was he that the terrestrial spirit group itself is sometimes referred to as the Agasha spirit group. Agasha was the name that Jesus went under when he was incarnated in Atlantis approximately 10,000 years ago and he was incarnated again 7–8,000 years ago in India where he was known as Krishna. Four thousand years ago he was seen again on Earth, this time in Egypt where he was called Clario. Even when he is living in the Real World, he never ceases to offer advice to those of us living here.

Shakyamuni's activities, which are based around the Law, can be likened to the brain and nervous system of the body and at the same time are like the network of veins that travel to the extremities. On the other hand, Jesus' work is based around love, and if Shakyamuni creates the veins, Jesus acts like the heart, supplying the blood and pumping it throughout the body. Without a heart, the various parts of the body would cease to function and, likewise, without the work of Jesus, the members of the terrestrial spirit group would fall into discord or strife and no longer be able to carry out their duties. Thanks to Jesus' work in supplying the whole of humankind with love, people have come to realize the necessity of loving each other. He is the embodiment of mutual love, the great power of unity, and has continued in his task for hundreds of millions of years. Another manifestation of Jesus' love can be seen in the medical spirit group that is one of the many powerful spiritual groups under his guidance. The number of spirits associated with the white light of love is extremely high, due, in part, to the frequency with which Jesus came down to Earth to teach. Not only that, it is an actual fact that a large percentage of the spirits in the Real World belong to this group, putting his teachings into practice.

Among the more powerful of Jesus' followers are the seven Archangels. These beings originally accompanied the Great Spirit, Enlil, when he brought a large number of colonists to Earth. But, since then, they have worked mainly with Jesus as his assistants. The names of these Archangels are: Michael, Gabriel, Raphael, Laguel, Saliel, Uriel, and Panuel—who replaced Lucifel after he was banished to hell. As the leader of the Archangels, Michael instructs and organizes people and has also been granted the power to halt the workings of Satan and his followers. Gabriel is in charge of communication and has played important roles in numerous cultures and civilizations. Raphael is responsible for ensuring the flow of love within art, while Saliel is leader of the spirits of healing, that is to say he carries out the practical teachings of Jesus, working to do away with disease. He appears in the Buddhist pantheon in the guise of Yakushi-Nyorai (the Nyorai the master of medicine). He was recently incarnated on Earth where he went under the name of Edgar Cayce. Uriel is the angel who is principally active in the field of politics.

## 8. The Work of Confucius

Of the other great spirits, one that stands out is Confucius. Born in China in his last incarnation, he is principally the god of learning. Knowledge is like water, it flows from high to low, and so it can be argued that Confucius is also in charge of order. Order represents one way of achieving harmony. I have stated that we have two objectives in training our souls, these being progress and harmony, and, to this extent, order is a very important element in attaining harmony. Order refers to the relationship between authority and obedience, dominance and submission. Confucius has been working to create an order, in which those close to Buddha stand at the top while those who are not remain beneath, in accordance with the will of Buddha. To put it another way, Confucius' main interest has lain in creating an orderly world through study and through the path of virtue to produce a symmetry that reflects the will of Buddha.

To sum up the roles of these great beings: Shakyamuni acts as the leader, creating a network of blood vessels throughout humankind; Jesus is like the heart, pumping blood through the veins; and Confucius creates order, facilitating harmony in relationships. If we look back over history we can see that Confucius has contributed to the creation of well-regulated societies. In the Real World, too, there are superior and inferior spirits, each living harmoniously in the place allotted to them. It can be said that Confucius played a major role in the creation of this orderly structure.

Confucius was also incarnated in Atlantis more than 10,000 years ago where he was known as Osiris. Several thousand years after that, he gave instruction from heaven, in the name of Osiris, to the Guiding Spirits of Light who emerged in Egypt to create the Egyptian civilization. He then taught the difference between right and wrong, between superior and inferior, and this led to the legend of Osiris's scale—in murals inside pyramids we see him using a scale to separate the spirits of people into good and evil.

### 9. The Work of Moses

In the previous sections I have explained the relative positions of Shakyamuni, Jesus, and Confucius, but there remains another famous spirit of the ninth dimension, the person who led the Israelites out of Egypt, Moses. He ranks along with Jesus and Confucius and is primarily responsible for controlling miracles.

There are numerous ways of demonstrating the power of Buddha, and one of the most effective is through the use of miracles. When any phenomenon that cannot be explained logically occurs, people who witness it feel the working of Buddha. For instance, when Moses divided the Red Sea to enable the Israelites to escape, or when he received light from heaven to engrave the Ten Commandments in stone, the people were overawed by the power these acts demonstrated. The color of light that governs miracles is red and it is under the command of Moses.

At present, Shakyamuni is leading the work of creating new cultures and civilizations while Jesus is governing the Real World in his stead. Confucius is in the throes of producing a new plan for the future of the Earth as a part of the universe, and of planning the evolution of humankind and the roles they should play in the universe. Then, what is Moses doing? Moses has been given the task of organizing the dissolution of hell, which has been in existence now for more than one hundred million years.

### 10. And So to the World of Planetary Consciousnesses

The ten Great Guiding Spirits carry out their work mainly in the ninth dimension, but where does this light that is split into seven colors come from? The truth is that it comes from the tenth dimension. The tenth dimension is the realm of the Planetary

Consciousnesses. Unlike the inhabitants of the other realms up to the ninth dimension, these spirits do not possess any of the attributes of humanity and have never appeared on Earth in human form.

There are three Planetary Consciousnesses who exist in Earth's tenth dimension. First, there is the Grand Sun Consciousness, which commands the positive principle and promotes evolution on Earth. Next comes the Moon Consciousness, which commands elegance, artistic beauty, grace, and the passive principle. In this way, the Grand Sun and Moon Consciousnesses combine to produce the duality of positive (yang) and passive (yin) principles that make up the world as we know it. It is thanks to the influence of the Moon Consciousness that the world is blessed with the passive elements that would otherwise be missing in a totally positive world. Femininity in contrast to masculinity, shade in contrast to light, night in contrast to day, the sea in contrast to mountains; it is Moon Consciousness's task to add these gentle touches to the composition of the world.

The third being which exists on the tenth dimension is the Earth Consciousness, which has nurtured the planet and cared for all the living things on it for the last 4.6 billion years. It is the life force of the planet Earth itself, the force that nurtures all things. It made the mountains, controls volcanoes, and orders the continental drift and other changes in the Earth's crust. It promotes the growth of vegetation and the propagation of animals.

These three beings have nurtured the planet we live on for countless eons and have a boundless influence on every aspect of its development.

Above the tenth dimension lies the eleventh dimension, where the Solar System Consciousness dwells. This being is one of the Stellar Consciousnesses. In the twelfth dimension there exists the Galactic Consciousness while beyond this in the thirteenth dimension is the Cosmic Consciousness of the universe. Beyond this, there exists the Primordial Buddha who is so far removed from us that it is entirely beyond the scope of our comprehension.

We humans are blessed with an eternal objective, and we are progressing toward infinite evolution, while at the same time working to actualize both progress and harmony. This is the truth about the world that surrounds humankind, showing the signposts for us to follow and the objective for which we all must aim.

In this work, *The Laws of Eternity*, I have focused on the world from the fourth to the ninth dimensions. I have explained that this is the true structure of the world and that the three-dimensional world in which we are presently living is not everything. I have said that our true nature is spiritual, and that as spirits we live in the worlds I have described. I sincerely hope that you will henceforth base your lives on this knowledge of the Truth and use it to give you the courage to live a fuller life.

# Postscript

Since this book was first published (in Japanese) in 1987, the Institute for Research in Human Happiness has experienced miraculous growth and I believe the reason for its unprecedented development lies in the scale of the Law it preaches.

This book contains only the Truth, the Absolute Truth, and demonstrates that, as a disseminator of this, I myself am the personification of Truth. The truth that this book is in fact "The Laws of Eternity" will be proven in the future by how widely the Laws of Buddha's Truth are accepted and how far and long they are conveyed. Those people who are spiritually aware will realize that the knowledge I reveal in this book has its origins in the ninth dimension.

If we were to compare the teachings of Zen Buddhism to a decorative hill that has been constructed in a garden, then the knowledge contained in this book is taller than Mount Everest. This book represents the secret treasure of humankind and it is pervaded by the immense compassion of El Cantare for humanity today.

*Ryuho Okawa*
*July 1997, for Japanese edition*

## What Is IRH?

The Institute for Research in Human Happiness (IRH) is an organization of people who aim to cultivate their souls and deepen their wisdom. The teachings of IRH are based on the spirit of Buddhism. The two main pillars are the attainment of spiritual wisdom and the practice of "love that gives."

*Keep updated with*
## IRH MONTHLY

featuring a lecture by Ryuho Okawa. Each volume also includes a question-and-answer session with Ryuho Okawa on real life problems.

For more information, please contact local offices of IRH.

*Also available*
## MEDITATION RETREATS

Educational opportunities are provided to people who wish to seek the path of Truth. The Institute organizes meditation retreats for English speakers in Japan and other countries.

# THE INSTITUTE FOR RESEARCH IN HUMAN HAPPINESS
### Kofuku-no-Kagaku

**Tokyo**
1-2-38 Higashi Gotanda
Shinagawa-ku
Tokyo 141-0022
Japan
Tel: 81-3-5793-1729
Fax: 81-3-5793-1739
Email: JDA02377@nifty.com
www.irhpress.co.jp

**New York**
2nd Fl. Oak Tree Center
2024 Center Avenue
Fort Lee, NJ 07024
U.S.A.
Tel: 1-201-461-7715
Fax: 1-201-461-7278

**Los Angeles**
Suite 104
3848 Carson Street
Torrance, CA 90503
U.S.A.
Tel: 1-310-543-9887
Fax: 1-310-543-9447

**San Francisco**
1291 5th Ave.
Belmont, CA 94002
U.S.A.
Tel / Fax: 1-650-802-9873

**Hawaii**
419 South St. #101
Honolulu, HI 96813
U.S.A.
Tel: 1-808-587-7731
Fax: 1-808-587-7730

**Toronto**
484 Ravineview Way
Oakville, Ontario L6H 6S8
Canada
Tel / Fax: 1-905-257-3677

**London**
65 Wentworth Avenue
Finchley, London N3 1YN
United Kingdom
Tel : 44-20-8346-4753
Fax: 44-20-8343-4933

**Sao Paulo**
(Ciencia da Felicidade do
Brasil)
Rua Gandavo
363 Vila Mariana
Sao Paulo, CEP 04023-001
Brazil
Tel: 55-11-5574-0054
Fax: 55-11-5574-8164

**Seoul**
178-6 Songbuk-Dong
Songbuk-ku, Seoul
Korea
Tel: 82-2-762-1384
Fax: 82-2-762-4438

**Melbourne**
P.O.Box 429 Elsternwick
VIC 3185
Australia
Tel / Fax: 61-3-9503-0170

# About the Author

Ryuho Okawa, founder and spiritual leader of the Institute for Research in Human Happiness (IRH), has devoted his life to the exploration of the spirit world and ways to human happiness.

He was born in 1956 in Tokushima, Japan. After graduating from the University of Tokyo, he joined a major Tokyo based trading house and studied international finance at the City University of New York. In 1986, he renounced his business career and established IRH.

He has been designing IRH spiritual workshops for people from all walks of life, from teenagers to business executives. He is known for his wisdom, compassion and commitment to educating people to think and act in spiritual and religious ways.

The members of IRH follow the path he teaches, ministering to people who need help by spreading his teachings.

He is the author of many books and periodicals, including *The Laws of the Sun*, *The Golden Laws*, *The Laws of Eternity*, and *The Starting Point of Happiness*. He has also produced successful feature length films (including animations) based on his works.